A. E. (Amos Emerson) Dolbear

First Principles of Natural Philosophy

A. E. (Amos Emerson) Dolbear

First Principles of Natural Philosophy

ISBN/EAN: 9783744749985

Printed in Europe, USA, Canada, Australia, Japan

Cover: Foto ©Thomas Meinert / pixelio.de

More available books at **www.hansebooks.com**

FIRST PRINCIPLES

OF

NATURAL PHILOSOPHY

BY

A. E. DOLBEAR, M.E., PH.D.

PROFESSOR OF PHYSICS AND ASTRONOMY, TUFTS COLLEGE, MASS., AUTHOR OF
"THE ART OF PROJECTING," "THE SPEAKING TELEPHONE,"
"MATTER, ETHER, AND MOTION"

BOSTON, U.S.A., AND LONDON
GINN & COMPANY, PUBLISHERS
The Athenæum Press
1897

PREFACE.

THE growth of physical science has rendered it more and more certain that phenomena of all kinds and in all places are due to the qualities and activities of the ultimate atoms of matter.

Astronomy, Geology, Chemistry, and Physiology are each easily reducible to the same factors, and hence these sciences may properly be classed as departments of physical science. The name Physics has heretofore been held to apply to phenomena which could not properly be included in the above sciences. For that reason, and to emphasize the relationship of these to fundamental physical principles, the name Natural Philosophy has been adopted for this book, restoring an old term to its proper place, and giving it its proper meaning.

The effort has been made to direct the attention of the student from the physics of mechanism to the physics of molecules, and help him to carry the mechanical conceptions gained by the study of visible bodies to their ultimate particles. The student is thus assured that molecular phenomena can in this way be accounted for without assuming other and different factors, and realizes that there is nothing more mysterious in the one than in the other. Thus the nature of the so-called forms of energy, and what it is that happens when energy is transformed, are made clear to mechanically minded persons.

In the attempt at simplification, some changes have been introduced in the treatment of energy and work. *Pressure* has been substituted for *force*, as a factor in phenomena, for the reason that if the latter be used in any other sense than as a pressure, it conveys no mechanical idea. At best it is an indefinite conception disguised in a mathematical *f*, and whatever advantage that may have for advanced thinkers, it has none for beginners.

The ether has suddenly become highly important for the proper understanding of the phenomena of magnetism, electricity, and light, and discoveries lately made have rendered it needful to change radically both theories and conceptions, and to restate nearly the whole of these subjects. These changes have been incorporated in the work in a way which it is hoped will commend itself to teachers and be found easy of apprehension by students.

The system of weights and measures in common use has been adhered to throughout, because there can be no confusion of mind in their use; also because no time will need to be taken from the study itself to learn to think in and apply a new system of units; and lastly, because the metric system is used nowhere except in laboratories, and not more than one in a thousand of those who will study Natural Philosophy will have occasion ever afterwards to use that system.

No space has been given to the history of the science, but it is to be hoped that every teacher of the subject will add to the interest of it by supplementing the lessons with stories of the lives, efforts, methods, and successes of the men eminent in all the fields of physical science. Nothing of so much importance can take the place of this, but it should come from the teacher rather than from an elementary text-book.

A. E. DOLBEAR.

CONTENTS.

NATURAL PHILOSOPHY.

CHAPTER I.

MATTER AND SOME OF ITS PROPERTIES.

Physical Science has to do with the properties of matter, its behavior under all conditions, and also the circumstances under which any kind of a change takes place in it, whether it be one of place, of form, or of condition.

It is sometimes described as the science of *Energy*, because it has been found that all kinds of changes in matter are due to energy acting in certain definite ways, which, when known, enable one to foresee what will happen, and often to use such knowledge for convenience, or business, or pleasure.

Every kind of an action that can affect in any degree any of our senses is called a physical action. For example, this book can be felt by the sense of touch; it can be seen by the sense of sight; if dropped upon the table, it can be heard by the sense of hearing. A peppermint lozenge can be felt, seen, heard, smelled, and tasted, and so reach five of our senses by five different kinds of physical action. It can be felt because it has body, it can be seen because light is reflected from it, it can be heard because it strikes another body, it can be smelled because some of its material evaporates and its particles reach the nerves of smell, and it can be tasted because some of its particles dissolve upon the nerves of taste upon the tongue.

Each one of these actions, and every other action capable of affecting us in any manner, is called a *phenomenon*, so physical science may be called the Science of Phenomena.

Matter. — Every kind of an object that can affect our senses either directly or indirectly is called *matter*. If the object be near at hand we may touch it; if farther off we see it, and find by experience, if we go to it, we may also touch it. Everything that shines or reflects light is matter. Everything that has form or color or brightness or taste or smell or weight is properly called matter. If matter be rare, like hydrogen, to feel it may be difficult, but it may be condensed to a liquid, and then be sensible to the touch and sight.

QUESTIONS.

1. Name some objects and state which of the senses they affect.
2. Describe some phenomenon.
3. Would you call the sound of a bell a phenomenon?
4. Is a star a phenomenon?
5. Can you name some object composed of matter?
6. Can you name some object not composed of matter?
7. How does one become aware of the existence of matter?
8. What reason have you for thinking that a house you have never been near is composed of matter?
9. Why do you infer that the sun and moon and stars are made of matter?
10. Can you think of some sensation not due to matter?

Kinds of Matter. — By subjecting substances to chemical action, such as dissolving them in acids or burning them in fire, it has been found that most of them are capable of being broken up into two or more constituents, thus showing them to be compound bodies. This is the case with wood, stones, earth, water, etc. These ultimate constituent parts are called *Elements.*

About seventy different elements are now known. Some of them all persons are familiar with, such as gold, silver, nickel, copper, iron, lead, tin, zinc, carbon. Other elements less familiar are antimony, silicon, calcium, bismuth, aluminum, oxygen, hydrogen, nitrogen; some are exceedingly rare, such as didymium, gallium, lanthanum. Each of these elements differs from every other in its physical and chemical properties, so it may be identified and separated from every combination containing it. Gold is yellow, copper red, and silver white. Some enter into chemical combination very easily, such as iron and oxygen, forming a mineral that is so plentiful as to form great hills and beds; it is called iron ore. Others will not enter into combination under ordinary conditions. For instance, carbon may be kept for an indefinite time exposed to air, water, or acids, and not be affected by them. There is good reason for thinking that most of the body of the earth is composed of iron; but the rocks that make up most of the land surface are composed of *aluminum*, *oxygen*, *silicon*, *calcium*, and *carbon*, while the water is a compound of *oxygen* and *hydrogen.*

QUESTIONS.

1. Name some of the common kinds of matter you chance to be familiar with.
2. Do you know whether they are elements or compounds?
3. How many of the elements have you ever seen?
4. How do those you have seen differ from each other?
5. How could you find out the composition of a substance?
6. How would you know lead from tin?

Divisibility of Matter. — A lump of salt may be pounded into dust so fine that a particle could be seen only by looking through a microscope, which might show it to be no bigger than the hundred-thousandth of an inch; but if the lump of salt be put into water the smallest particle produced by pounding will be dissolved into more minute parts. Yet this process cannot go on without limit. There is a smallest part of salt, which, if broken up into pieces, is no longer salt, but the elements which were united to form the salt, — namely sodium and chlorine. The smallest part into which any substance may be divided without destroying its qualities is called a *molecule*, while the constituents of the molecule are called *atoms*. Thus a molecule of salt is composed of an atom of sodium combined with an atom of chlorine, and is chemically written $NaCl$. A molecule of water is composed of two atoms of hydrogen and one atom of oxygen, and is written H_2O. Chemistry is the science that treats of the combination of atoms into molecules, and the proper conditions for combining or separating substances in order to produce any desired kind of matter, either simple or

LIST OF COMMON ELEMENTS AND THEIR CHEMICAL SYMBOLS.

NAME OF ELEMENT.	SYMBOL.	AT. WT.	DENSITY.	HARDNESS.	MELT. POINT.
Aluminum	Al	27	2.6	3	1260
Antimony	Sb	120	6.7	3	790
Arsenic	As	75	5.7	3	—
Bismuth	Bi	208	9.8	2	522
Bromine	Br	80	3.1	—	20
Calcium	Ca	40	1.6	1	1200
Carbon	C	12	2	1.10	—
Chlorine	Cl	35.5	1.3	—	—
Copper	Cu	63.2	8.9	3	1900
Gold	Au	196	19.3	3	1900
Hydrogen	H	1	—	—	—
Iodine	I	126	5	—	230
Iron	Fe	56	7.8	5	2900
Lead	Pb	207	11.3	2	600
Magnesium	Mg	24	1.7	2	1170
Mercury	Hg	200	13.6	—	—
Nickel	Ni	58	8.9	4	2700
Nitrogen	N	14	—	—	—
Oxygen	O	16	—	—	—
Phosphorus	P	31	1.8	—	110
Platinum	Pt	195	21.5	4	3400
Potassium	K	39	.87	—	140
Silicon	Si	28	2.3	—	—
Silver	Ag	108	10.5	2	1800
Sodium	Na	23	.98	—	200
Sulphur	S	32	2	2	235
Tin	Sn	118	7.2	2	440
Zinc	Zn	65	7	3	780

compound. Under ordinary conditions of temperature most, if not all, of the elementary atoms combine with their like if there be no other kind present with which they can unite: hydrogen unites with hydrogen, one atom of each, H,H; oxygen, O,O; carbon, C,C; and so on; and special means have to be applied to obtain matter in its atomic form. Hence iron, carbon, oxygen, sulphur, are molecular in constitution, but of similar atoms, while compound substances are molecular but of dissimilar atoms.

Some molecules consist of but two or three atoms, as NaCl, H_2O, while others may have tens and hundreds of atoms. For example, a molecule of albumen from the white of an egg consists of $C_{210}H_{330}N_{52}O_{66}S_3$, 661 atoms. It must be evident that molecules vary much in size. Gold may be beaten out into leaves less than the millionth of an inch thick; and it must be more than one molecule thick, or it could not hold together. The tiniest bit of silver dissolved in nitric acid will render milky a hundred cubic inches of a solution of common salt. If the silver be the .001 of an inch cube, it will have been divided into $1000^3 \times 100 =$ 100000,000000 parts.

Drop a bit of aniline the size of a pin's head into a quart of water. It will give decided color to it. Calculate how many times the volume of water is greater than that of the aniline, to find how many parts the latter has been divided into. Half fill a test tube with the colored liquid and fill up with water; this will dilute it one-half and double the number of divisions. This may be continued until the color ceases to be discernible. Still finer divisions may be perceived by dissolving in a similar way eosin and resorcin.

Size of Molecules.—Blow a soap-bubble and observe the bands of color that flow down the sides. After the last purplish tint comes a white band; and following this is a dark gray patch from which but little light is reflected. The thickness of this gray patch has been shown to be but the $\frac{1}{160}$ of the wave-length of the purple light, which is in the neighborhood of $\frac{1}{60000}$ of an inch. $\frac{1}{60000} \times \frac{1}{160} = \frac{1}{9,600000}$, hence the thickness of the film of the bubble cannot be far from one ten-millionth of an inch. Yet it possesses tensile strength enough to hold up the weight of the bubble, and therefore is greater probably than one molecule thick; if it be two molecules thick their diameters will be but one twenty-millionth of an inch, and if the film be five molecules thick, each must be about the fifty-millionth of an inch in diameter. Various phenomena indicate that the size of molecules of two or three atoms is from the fifty-millionth to the hundred-millionth of an inch, the atoms being correspondingly smaller. For most purposes it will be convenient to bear in mind that about fifty million molecules in a row would reach an inch, and the number in a cubic inch, say of water or iron, would be the cube of fifty millions, 125000,000000,000000,000000. (Where such large numbers occur it is customary to write the significant figures multiplied by ten raised to the power indicated by the number of ciphers. Thus the above would be 125×10^{21}.)

Molecules with a large number of atoms must of course be correspondingly larger, so that the diameter of a molecule of albumen, containing over 600 atoms, would probably be eight or nine times that of water, that is, about the millionth of an inch. As the highest power of the microscope enables one to see a particle but about the hundred-thousandth of an inch in diameter, it is clear it could not make visible a molecule even of the largest size, and the smallest particle visible would contain not less than a hundred million atoms. If one would gain something of an idea what a prodigious number of atoms there are in a cubic inch of ordinary matter, let him compute how long a time it would take to count them at any definite rate, say one thousand or a million a second.

QUESTIONS.

1. What reason is there for thinking that there is a limit to the divisibility of matter?

2. What is the difference between the sciences of physics and chemistry?

3. What is the distinction between an atom and a molecule?

4. Are the elements ordinarily found in the atomic or in the molecular state?

5. How many atoms are there in a molecule?

6. Can you find in any book on chemistry a molecule with more atoms than there are in albumen?

7. How many molecules of albumen will there be in a cubic inch if each atom in the molecule be the fifty-millionth of an inch in diameter?

8. If a grain of aniline will give color to a cubic foot of water weighing 62½ pounds, into how many parts has the aniline been divided?

9. Can you think of any reason why that number does not represent the limit of division of the aniline?

Shape of Atoms. — It was formerly thought that atoms must be minute solid spherical particles, so hard they could not wear out, but now it is believed that such kinds of atoms could not possibly present such phenomena as matter actually possesses, and there is an increasing probability that atoms are minute vortex rings, similar in form to such rings as are sometimes puffed out from a locomotive in still air, where they may be seen to rise a hundred or more feet high, vibrating and wriggling in a curious way, but maintaining their ring form until dissolved in the air. Such rings[1] possess form, rigidity, elasticity, inertia, and energy. If it were not for the friction in the air, when once formed

FIG. 1.

[1] Vortex rings for illustration may be made by having a wooden box about a foot on a side, with a round orifice in the middle of one side, and the side opposite covered with stout cloth tightly stretched over a framework. A saucer containing

FIG. 2.

strong ammonia water, and another containing strong hydrochloric acid, will cause dense fumes in the box, and a tap with the hand upon the cloth back will force out a ring from the orifice. These may be made to follow and strike each

they would be practically indestructible things, possessing on a large scale such properties as atoms exhibit on a minute scale. Atomic vortex rings are supposed to be constituted of ether,[1] which is known to be frictionless. There is now no other theory of atoms than this one, and it does not appear that there are any serious objections to it. For that reason such atoms will be assumed throughout this book.

QUESTIONS.

1. Take up a small drop of water on the point of a pin. Estimate its size, and compute how many molecules of water make it up.

2. A blood-corpuscle is about the three-thousandth of an inch in diameter and one-fifth as thick; how many such can there be in a cubic inch of blood?

3. How many molecules are there in a cubic inch of air if they are two-hundred-thousandth of an inch apart on the average?

4. If the space of a cubic inch were made a perfect vacuum, and a minute hole made into it, and ten million molecules of air a second were to go through, how long would it take to be filled?

States of Matter. — As the atoms and molecules are much too small to be seen, what we do see is made up of immense numbers of them compacted together. If the molecules cohere so strongly in fixed positions that they do not fall apart easily we call the body a *solid* body. A book or pencil or stone is an example of a solid; and there are all degrees of compact-

other, rebounding and vibrating, apparently attracting each other and being attracted by neighboring bodies.

By filling the mouth with smoke and pursing the lips as if to make the sound *O* one may make fifteen or twenty small rings by snapping the cheek with the finger.

[1] See page 138.

ness or solidity. When molecules have but slight attraction for each other, and may be separated very easily, the body is called a *liquid*. The molecules of water cohere slightly, else they would not be formed into a drop. When water is poured into a pitcher or other vessel it immediately assumes the shape of the containing vessel; for a liquid has no particular form, except that a small quantity, like a drop, will assume a spherical form if free in the air, and a spheroidal form when resting upon a surface it cannot wet. Drop gently a little water upon a painted or varnished surface and notice the rounded edges and flattened top it assumes. The smaller the drop the more nearly it is spherical. The same may be seen still better by using a little mercury instead of water. This shows slight molecular attraction. Indeed it appears as if the molecules can move freely among themselves, and do not cohere in any *fixed positions*.

The molecules of some substances do not cohere at all, but act as if each one endeavored to get as far from the rest as it could; and being free to move each bumps against its neighbors and rebounds from them, so that a quantity of such free roving molecules has no particular shape under any conditions, but practically fills all the space of any vessel in which it may be inclosed, as a swarm of flies might do. Such free roving molecules without cohesion is called a *gas*, and a single one is called a gaseous molecule. The air we breathe is an example. If a bottle of ammonia water be opened, the ammonia gas will escape, and presently may be detected throughout the room. Odors of all sorts are gases, whose individual molecules are free to move in every direction. One may imagine a single molecule of any kind in a room. Now it may be here and now there. It would not be possible to detect a single one, but when numerous enough some may be detected by smell, as cologne; some by chemical methods. Remove the stoppers from a bottle of hydrochloric acid and one of ammonia when they stand a foot or two apart, and a white cloud will be formed in the air between them. If any gases are condensed so as to bring their molecules into approximate contact and held there, they form a liquid. Such condensation of gases may be effected by cold and pressure; and so ammonia, air, oxygen, and indeed all gases have been reduced to liquids, and some even to the solid form, as in the case of carbonic acid gas. A solid will maintain its *shape* and volume. A liquid assumes the shape of the containing vessel while its volume remains the same; but a gas keeps neither shape nor volume, except when confined.

These three states of matter depend almost altogether upon temperature. For instance below 32° the liquid water becomes ice, a solid, and above 212° it becomes steam, a gas. Below —39° mercury becomes a solid, and at 357° it is a gas. Iron, platinum, and all the solid metals are converted into gases by the heat of an electric arc; while earth, stones, and such solids are either fused to liquids, like the lava of volcanoes, or are vaporized, which is but another name for assuming the gaseous form. At the sun it is probably so hot that there are no solids. At the moon it is probably so cold that there are no liquids or gases.

At the earth's surface the air particles have a free path to move in, between collisions, over two hundred times their diameter. If a molecule were magnified to the size of a honey-bee, then the corresponding space for free movement would be about ten feet without coming in contact with another molecule. As one ascends above the surface of the earth the air is less and less dense, — that is, there is a smaller number of molecules to the cubic inch, and each molecule has more and more free space to move in, a longer *free path.* At the height of 200 miles a molecule might have a free path of 50,000,000 miles without colliding with another.

Porosity. — As there is so great a distance between the molecules of common air, it is plain that there is room for others of any kind that may chance to be present. Indeed, that molecules of ammonia or other gas become diffused through the air is evidence that the air particles do not occupy all the space. But a liquid differs from a gas in having its molecules closer together; but even then they do not occupy all the space. Salt, sugar, and many other things will dissolve in water without increasing its bulk appreciably, which shows that there are spaces between water molecules where other molecules may find place. The same is true of solids. Gold and silver will dissolve in mercury nearly as freely as salt in water, and a drop of mercury will be absorbed by most metals like water into wood. This fact, that solids and liquids are able to share their space with other substances, is called *porosity*, and is due to the fact that the molecules of all substances, even the densest, are not in such contact as to take up all the space. If the molecules were shaped like marbles there would be inter-molecular spaces. Fill a cup with marbles or shot and a good deal of water can then be poured into the cup. When it is apparently filled with water, a good deal of sugar can be added. If the atoms are ring-shaped, they must still leave a good deal of unoccupied space. Water can be squeezed through iron an inch thick, and gases through gold and platinum. The pressure of the wind is quite sufficient to force air in a steady stream through brick and plastered walls a foot thick.

QUESTIONS.

1. How do solids differ from liquids?
2. Whittle small flat surfaces on two bits of lead, and press the surfaces together with a slight twisting movement. Note what happens, and describe it.
3. Can you do the same for wood? If not, why not?
4. How does it happen that gaseous molecules fill any kind of a vessel they chance to be in?
5. Is it true for all quantities of gas? Why?
6. What idea have you of a gas and why it behaves as it does?
7. What causes the difference between ice, water, and steam?
8. Can you imagine how it does it?
9. Name substances you know to be porous. How do you know them to be so?
10. What reason is there for thinking that water is porous?
11. If matter *filled* space would it be porous?

Density. — By this is meant the amount of matter in a given space. If two cubic inches of air are made to occupy the space of one cubic inch, then there is twice as much air in that space as before, and we say the density is twice as great. The number of molecules per cubic inch determines the density, but it would be inconvenient to employ such a number if it could be accurately found. We use instead the weight of bodies to determine their density. Water is taken as a standard for solids and liquids. A cubic foot of water weighs 62½ pounds. If a cubic foot of iron weighs 468 pounds, we say that the density of iron is 7.5, for $\frac{468}{62.5} = 7.5.$

In the Table of Elements on page 5 is a column giving the relative densities of the elements. The numbers signify how many times heavier the substance is than an equal volume of water. Thus, the density of aluminum is 2.6; gold, 19.3. The numbers also represent the specific gravity of the different elements; and how they may be determined is shown on page 65.

For gases, either hydrogen or common air is used as a standard; 100 cubic inches of air weigh 31 grains. [When hydrogen is employed, the standard volume is a liter (1000 cubic centimeters), which, at standard temperature (32°) and pressure (30 inches of mercury), weighs 0.0896 of a gram.]

QUESTIONS.

1. A cubic foot of marble weighs 168 pounds; what is its density?

2. If the density of lead be 11.3, how much will a cubic foot weigh?

3. What must a bar of iron weigh that is an inch square and a foot long?

4. How much does a cubic foot of air weigh?
5. What volume of air weighs 100 pounds?
6. How much does a cubic foot of gold weigh?
7. If a gallon of alcohol weighs 50 pounds, what is its density?
8. If the density of the earth be 5.6, how many tons does it weigh?

Elasticity. — When a piece of India rubber is stretched and then released it returns to its original length. Likewise a rubber ball, if pinched, will be deformed, but it will regain its spherical shape as soon as the pressure is removed. This ability of a body to recover its original form after distortion is called *elasticity*, and all bodies possess it in some degree, but with great differences. Glass, ivory, and tempered steel are highly elastic. Putty, dough, and wet clay have but a slight degree of this property. All liquids are very

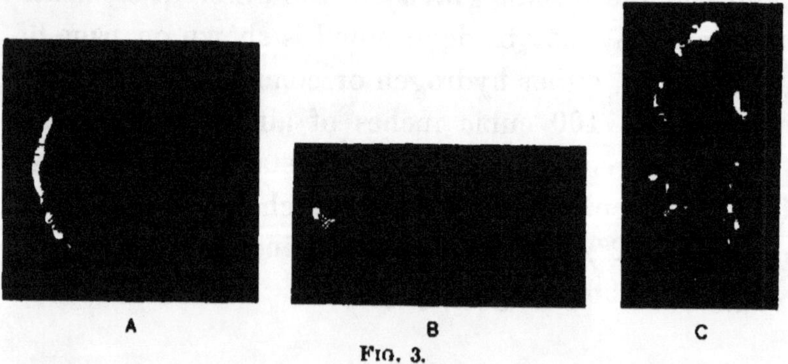

Fig. 3.

elastic. Water, if compressed, as it may be, will recover its original volume at once, no matter how long a time it has been kept under stress. Gases also possess elasticity, which depends upon their temperature, and there is reason for thinking that the ultimate atoms have this property as one of their essential ones. When vortex-rings collide they bound away from each other as billiard balls will do; and they may be seen to vibrate, their sides swinging to and fro like the prongs of a tuning-fork that has been struck (Fig. 3); and for a similar reason they endeavor to recover their original form.

A ring made of steel or brass wire 6 or 8 inches in diameter, if pulled out to an ellipse with thumb and

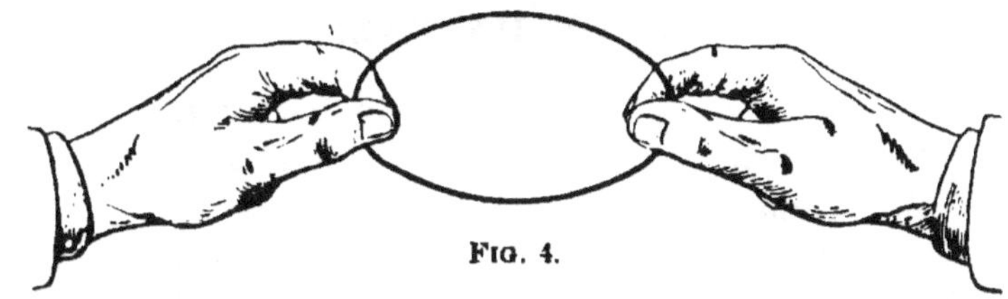

FIG. 4.

fingers and then let go, will vibrate well, and show this elasticity due to form.

Hardness. — A piece of pine wood can be cut easier than a piece of oak, and a piece of oak easier than a piece of brass, and brass easier than steel. There are all degrees of difficulty of this kind in substances of different sorts. We say one substance is harder than another when it is less easily cut or scratched; but when one cuts any substance he is separating the molecules of that substance and these cohere with different degrees of strength. The same attraction that causes atoms to combine into molecules causes molecules to adhere together. Only for convenience we speak of the former as chemical attraction or chemism, while the latter is called cohesion or adhesion; and the hardness of a body depends on how strongly the adjacent molecules stick together. The terms hard and soft do not apply to gases or liquids, because their molecules have little or no attraction for each other; only solids exhibit it, and its chief importance is as characteristic of minerals. A *scale of hardness* has been devised from talc, which is the softest of minerals, and may be cut without dulling a knife much, to the diamond, which is the hardest of known bodies.

SCALE OF HARDNESS.

1. Talc.	6. Feldspar.
2. Gypsum.	7. Quartz.
3. Calcspar.	8. Topaz.
4. Fluorspar.	9. Sapphire.
5. Apatite.	10. Diamond.

Each one of these in their order will scratch the one preceding it, and will not be scratched by it. Quartz will scratch feldspar, and the diamond will scratch sapphire; but only diamond-dust will scratch the diamond. But even among diamonds there is as great a variety in hardness as there is among other minerals. Some are so hard they cannot be scratched at all, and cannot be used as gems because they cannot be properly shaped. A body may be hard yet fragile, for a blow with a hammer that would not affect a piece of steel with a hardness of 7 would break a diamond into many fragments. Some of the metallic elements are so soft they may be moulded into any form in the fingers. Such is the case with sodium and potassium; also with selenium at a temperature of 140° F.

In the Table of Elements (page 5) is given a column containing the hardness of such as are solid.

Mass. — We determine the amount of matter in a given substance by weighing it. There is twice as much matter in two pounds of sugar or of iron as there is in one pound; and there is, therefore, as much matter in a pound of sugar as there is in the pound-weight that balances it on the scales. The amount of matter in a body, as determined by properly weighing it, and independent of its volume, is called its *mass.* A pound of feathers may occupy a cubic foot, or it may be condensed to a few cubic inches. In either case the number of molecules is the same, the weight the same, and therefore the mass is independent of the space the matter occupies. But the weight of a body is caused by the attraction of the earth, and such attraction varies with the distance from the surface of the earth. At the center of the earth a body would weigh nothing, because equally attracted on all sides, and therefore would not tend to move in any direction. At the distance of 10,000 miles from the earth, the attraction of the earth would be small upon it, and it would consequently have small weight. In either case the number of molecules would be the same, and the mass would be the same. There is another way of determining mass, without weighing, described on page 21. Atoms of the different elements have different masses. They are represented in the Table of Elements, page 5, by the numbers in the column marked "atomic weight," in which the mass of an atom of hydrogen is called unity.

Thus, the atom of oxygen has 16 times the mass of hydrogen; that is, it takes 16 atoms of hydrogen to weigh as much as one of oxygen. An iron atom has 56 times, and a gold atom 196 times the mass of a hydrogen atom.

Gravity. — That the earth attracts bodies so as to give to them weight has been known for hundreds of years. Newton discovered that *all* matter was thus attractive; that every atom of matter attracts every other atom in the universe, the strength of the attraction between two bodies depending upon the quantity or mass of matter and the distance between them. The law of gravitation is, every atom of matter attracts every other atom. The strength of the attraction varies with the mass of each atom, and is inversely as the square of their distance apart, that is, the attraction of two atoms of a given kind for another atom is twice as much as the attraction of one; and when the distance between two masses is doubled, their attraction is reduced to one-fourth. If the distance be tripled, their mutual pull upon each other will be only one-ninth, and so on. Gravitative attraction is very weak between masses of only a few pounds, and requires very delicate experiments to enable one to observe it. When one considers the enormous number of atoms that compose the earth, the moon, and the sun, he will see that the weak attraction of a one-pound mass is multiplied by thousands of millions. The attraction between the earth and the moon is equal to twenty quadrillion tons.

The strength of the earth's gravitative attraction is measured from its center, which is 4000 miles from its surface ; hence, a mass that weighs one pound at the surface of the earth would weigh but one-fourth of a pound 4000 miles above the earth, for the distance from the center is doubled, and at the distance of the moon, 240,000 miles (60×4000), only the thirty-six-hundredth of a pound $\left(\frac{1}{60^2}\right)$. At the surface of the sun the strength of gravitative attraction is 28 times its value upon the earth, so that a person weighing 150 pounds here would weigh there $150 \times 28 = 4200$ pounds, nearly two tons, and of course would be quite unable to move. The sun attracts every ton on the earth with a pull of a little more than a pound. An equal mass upon the moon is attracted by the earth with a pull of 10 ounces.

As the weight of a body depends upon the attraction of gravitation at the place where it chances to be, if the weight of the body be divided by the value of gravity, the quotient will be the same quantity at all times and all places. Thus, if the value of gravity on the earth be 32, and at the sun $28 \times 32 = 896$, a body weighing one pound on the earth would be represented by $\frac{1}{32}$. On the sun the same body would weigh 28, and would be represented by $\frac{1 \times 28}{896} = \frac{1}{32}$, precisely the same quantity.

Let w equal the weight of a body, and g the value of gravity where the body may be, then $\frac{w}{g} = a\ constant$

quantity; and this is sometimes treated as the mass of the body. If m = the mass, then one may write, $\frac{w}{g} = m$. The fraction $\frac{1}{32}$, as above, is sometimes called a *poundal.*

QUESTIONS.

1. If a ton were raised 4000 miles above the surface of the earth, what would it weigh there?
2. Would its mass be any less?
3. Do you think you could move it by pushing any easier than at the earth's surface?
4. How far away must it be removed in order to weigh one pound?
5. What is the difference between weight and mass?
6. How far from the earth must a body be moved that it shall weigh nothing?
7. Would it make any difference in which direction a body should be moved?
8. Would you think there might be a point between the moon and the earth where any mass of matter would weigh nothing?

CHAPTER II.

MOTION.

If all matter were quiescent there would be no changes anywhere, for, as most persons already know, all the changing phenomena that make life possible and interesting are due to movements of some kind. If the world should stop turning upon its axis there would be continuous day or night. If it should stop its motion about the sun there would be no change of seasons. If the molecules should stop vibrating there would be neither heat nor light, for such motions constitute heat and give rise to light, and in the absence of these all life would cease; indeed, the world would be a most uninviting place to be in, even if one could exist upon it. If, then, all kinds of phenomena are caused by motion, it is proper to study the characteristics of it in order to discover how so great a variety can flow from it.

Motion means change of position or of place. It is not very easy to give a definition that shall cover all cases. Two persons in a car-seat may be moving at the rate of 50 miles an hour, yet be no more disturbed by it than if they were in a house. If one walks backward upon a train as fast as the train moves forward, he may talk all the time with a man standing still by the side of the track. With reference to the standing man he might be still; with reference to the train he might be walking four miles an hour. Motion is therefore relative; that is, it depends upon what the moving body is compared with. If a man walks round a tree facing the tree all the time, he has faced every part of the horizon, he has turned completely round. If he walks round the tree facing in one direction all the time, he will have seen the tree on one side only. The moon constantly faces the earth with the same side, as if it were fixed to the end of a rod connected with the earth. It therefore turns upon its axis, although it appears not to.

Practically there is little or no difficulty in employing the term motion; a moving body may go fast or slowly, it may go in a straight or a curved line, to and fro, or round and round. If one imagines a point to have motion, its direction and velocity may be any assignable one. The consideration of these gives rise to a mathematical science called Kinematics.

KINDS OF MOTION.

I. Translational. — A body of any magnitude may move in any direction in free space, north or south, east or west, up or down, or in any intermediate direction. A body thus moving freely in any direction is said to move in a *free path*, no matter whether it move a long or a short distance. Nor does it matter what the size of the body may be: a cannon ball may go through the air eight or ten miles without striking anything; while a molecule of air may move no more than the millionth of an inch, before striking another molecule, yet this distance would be fifty times its own diameter. When a body moves as a whole from one place to another with such free-path motion it is said to have translatory motion ; if it goes in a straight line, like a billiard ball, the motion is called rectilinear; and if in a curved line, like that of a cannon ball, it is called curvilinear motion.

The drifting of clouds, the flight of birds, of arrows, the movement of meteors, comets, and planets in their orbits, are examples of translatory motion, and the student may think of and name others of the same kind.

II. Vibratory. — To and fro movements like the swinging of a clock pendulum (Fig. 5), the movement

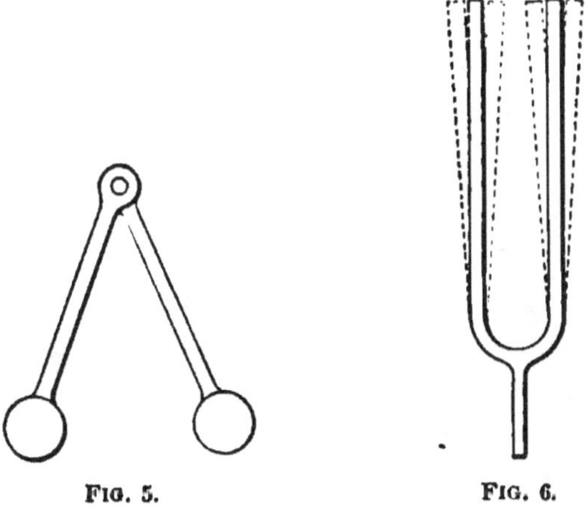

FIG. 5. FIG. 6.

of the piston of a steam-engine, the swaying of the branches of a tree, the vibrations of the prongs of a tuning-fork (Fig. 6), the reeds and strings of musical instruments, are examples of a different kind of motion, in which the changes of position are of the parts of a body with relation to itself rather than to other things. The prongs of a tuning-fork approach and recede from each other, each retracing its path — a characteristic of what is called vibratory motion. If the moving part be large, and its motion conspicuous, like the pendulum of a clock, the motion is sometimes called oscillatory.

III. Rotary. — A body of any size or shape may be made to turn upon some axis or spin around like a top or wheel. In such motion the parts of the body do not change their relative positions with reference to each other, but with reference to other bodies away from them. Thus, every part of a turning wheel is presented each revolution to opposite points in space. Such motions are called rotary; and examples of it are to be seen everywhere.

No one of these kinds of motion is peculiar to any body. They are applicable to masses of all magnitudes. An atom or a molecule may spin upon its axis as well as does the earth; and in general, *what a body will do depends upon what kind of motion it may have.* By spinning a base ball it may be made to go through the air in a curved line instead of a straight one; and a spinning top (Fig. 7) will stand on its point, which it will not do if it be not spinning.

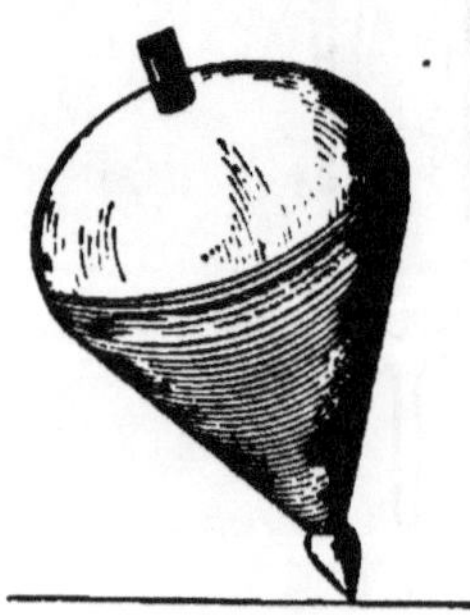

Fig. 7.

A vibrating tuning-fork or bell will apparently attract small bodies to itself; so from the behavior of a body one may often know what kind of motion it has without examining it directly.

These three kinds of motion — *translatory*, *vibratory*, and *rotary* — are the *primary motions*. All the various kinds of movements seen in machines are due to these and their compounds; and obviously a body may have any one or a combination of them. A bullet let fall goes in a translatory motion to the ground. If it be shot horizontally from a gun it will go in another translatory direction, which is a compound of horizontal and vertical translatory motions, that is, its curved path will be a compound of two translatory motions. In the diagrams on page 28 are represented these different kinds and their combinations.

I. A body at **A** may move on in the direction **AB** for an indefinite distance.

II. The body may move to and fro between **A** and **B**.

III. A point may rotate in the same path indefinitely.

IV. The resultant of two translatory motions **AB** and **AC** upon a given body at **A** will make it move in a straight line **AD**, if the velocities in the two directions be uniform.

V. If the velocity of one be increasing in one direction, as **AC**, the path taken will be curvilinear, as **AD**.

VI. A body moving back and forth over a line like **AB** in Fig. **II**, and given a forward motion at right angles to the vibratory movement, would make an undulatory path which would be a combination of vibratory and translational motion.

VII. The motion of a screw when worked into wood is an illustration of a combination of rotary and translatory motions; the rotary in twisting the screw, and the translatory in advancing into the wood. Also if the end of a rope 20 or 30 feet long be held in the hand, and the hand be shaken up and down, a series of waves will travel along the rope, composed of the vibratory and translatory motions. If the hand be swung round and round a spiral motion in the rope will be the resultant, as in **VII.**

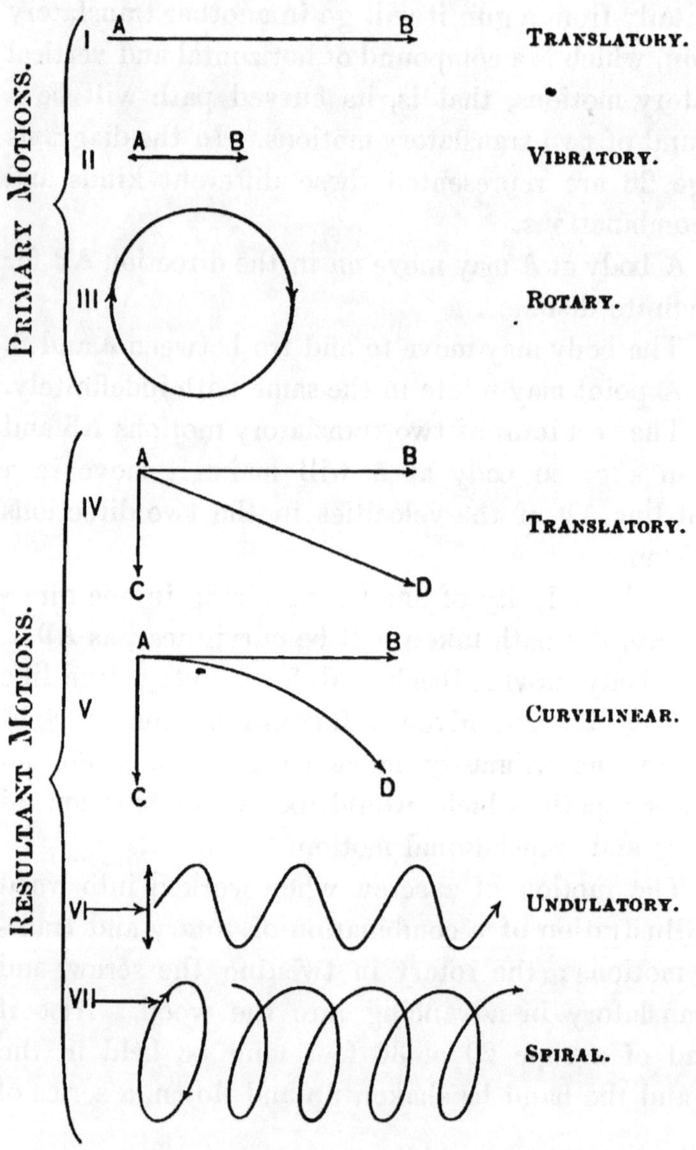

If one will observe the various kinds of movements that may be seen in a sewing-machine, he will be able to identify several of the foregoing: the vibratory motion of the treadle; the rotary of the balance-wheel, pulley, and spindle; the vibratory of the needle and shuttle; and translatory of the thread and the cloth. In like manner one should trace the motions in engines, looms, and other complicated machines, with the purpose of finding whether there be any other motions than the ones described here; and if so, how they are composed.

Rates of Motion. — The rate of change of position of a body is called its velocity; as when we say that a man walks at the rate of three miles an hour, or that the speed of a bullet is 1000 feet per second. These statements contain references to both space and time, and make it needful for us to use some standard for each. The unit of length that is most commonly used is the *foot;* if shorter spaces are being considered we use the inch, if long distances are considered we use the mile, which equals 5280 feet.

The standard of time used everywhere is the *second.* Practically we measure time in seconds, minutes, hours, days, and years, and whether we use one or the other depends upon what the special needs are. The year — 365¼ days — is the time it takes for the earth to complete one circuit about the sun. The day is the time it takes for it to turn on its axis once. The hour is the twenty-fourth part of the day, and the minute the sixtieth part of an hour. The second is the time it takes for a pendulum 39.117 inches long to vibrate once, and sixty such vibrations measure the time of one minute, and 86,400 of them are completed while the earth turns round once; that is, 86,400 seconds, is the measure of the length of the day.

The *rate of translatory motion* is measured as a distance per second, per minute, or per hour, but it is not to be understood that the motion is necessarily maintained through the time-period mentioned. For instance, when we say that the velocity of a bullet is 1000 feet per second, it means only that if the motion was maintained uniform for a second the bullet would travel 1000 feet. It might travel at the rate of a 1000 feet per second and go only a foot. A train of cars might go at the rate of 50 miles an hour and not travel a mile. Velocity is the rate of motion at a given instant. When the rate is known, and the time is given, the distance is found by multiplying the rate by the time.

Let v equal the rate and t the time, then the distance, $d = vt$. Any two of these being known, of course the other one can be found.

TABLE OF OBSERVED VELOCITIES.

	Miles per hour.
Man walking	3–4
Horse trotting	10–30
Gentle wind	6
Railway train	50
Hurricane	100

	Feet per second.
Man walking	6
Sailing vessel	15
Steamships	30
Crow flying	40
Race horse	42
Swallow flying	134
Swift flying	250
Sound in air	1100
Cannon ball	2000
Hydrogen molecule	6000
	Miles per second.
Shooting stars at earth	22
Comet near the sun	400
Light	186,400

The *rate of vibration* means the number of vibrations per second. Thus, a tuning-fork may vibrate 256 times a second; the wings of a duck 3 or 4 times; the wings of a mosquito 1000 times; and the chirrups of crickets represent about 3000 vibrations per second. Piano strings have a range from 40 to 4000. Steel rods have been made having a rate as high as 20,000 per second. In each of these cases a to and fro movement represents one vibration. In general, the smaller a body is the more rapid is its vibratory rate, provided it have the same form and be made of the same material; so the extremely minute atoms of matter have very high vibratory rates. A hydrogen atom, when made incandescent, vibrates not less than 450 millions of millions of times in one second. How this is known will be pointed out in the chapter on ether waves.

Sometimes it is convenient to know the actual space that is moved over by a vibrating body in terms of translatory motion. If the prong of a tuning-fork move the hundredth of an inch each vibration, and if it vibrates a hundred times per second, the total distance will be $100 \times \frac{1}{100} = 1$ inch. The extent of the swing of any point of a vibrating body on each side of its position when at rest is called the *amplitude;* and in the above case the amplitude is the four-hundredth of an inch. Suppose then the amplitude of vibration of a hydrogen atom be one-fourth of its diameter, or two hundred-millionths of an inch, and that it vibrate 500 millions of millions of times per second, the actual space moved through in that interval will be :

$$\frac{500,000000,000000}{100,000000} = 5,000000 \text{ inches} = 80 \text{ miles.}$$

Rotary Speeds have wide ranges, and are measured by the number of revolutions per second or per minute. A wheel may turn round fast or slowly; the balance-wheel of an engine may turn round once a second; high-speed steam-engines may turn four or five times per second; small wheels have been made to rotate 800 times a second. Sometimes, when great accuracy is required the velocity is measured in degrees per second. For instance, if a disk turns $5\frac{1}{2}$ times per second, a given point in its circumference moves at the rate of $360 \times 5\frac{1}{2} = 1980°$ per second.

Again, the rate of rotation may be considered as the time, t, it takes to make one revolution. The circumference of a circle of radius r is $2\pi r$ ($\pi = 3,1416$), and when that is described in time t, with velocity v, we have $2\pi r = vt$; v being the equivalent of translational motion.

A driving-wheel of a locomotive that is six feet in diameter will advance nearly 19 feet each revolution. If the locomotive is to move at the rate of a mile a minute, the wheel must turn round $\frac{88}{19} = 4.6$ times per second. A disk four inches in diameter, turning 800 times a second, would advance with a speed of $\frac{2\pi r \times 800}{12} = 837$ feet per second, nearly 10 miles a minute.

QUESTIONS.

1. If the distance to the sun be 93,000,000 miles, what is the velocity per second of the earth in its orbit?
2. If the earth be 8000 miles in diameter, what is the velocity per hour of a point on the equator due to the daily rotation?
3. If the velocity of sound be 1100 feet per second, how long will it take a sound to go quite round the earth?
4. If a duck fly 140 feet per second, how long will it take it to fly from Labrador to Florida?
5. A balance-wheel ten feet in diameter revolves $2\frac{1}{2}$ times a second; what is the velocity of a point upon the rim?
6. If a wheel make five rotations per second, what is its angular velocity?

Uniform and Varying Rates of Motion. — So far it has been assumed that the rate of motion of each kind was uniform, that is, it did not change during the interval of time considered. We know, however, that if a ball be thrown into the air it is soon stopped. It is the same with a bullet shot from a rifle, a top set spinning, a stone made to slide on glare ice, — indeed every kind of body having motion of any kind given to it and left to itself gradually comes to rest. We attribute to friction the loss of motion on the ground and in the air. Suppose, however, that there was nothing to interfere with the motion, that the body did not lose motion by imparting it to other bodies, how long would any body continue to move under such a condition? Evidently, if it did not give up its motion to anything else it would go on with the same velocity indefinitely, that is, its motion would be uniform.

A billiard ball goes in a straight line until it strikes something; it does not turn from its course unless compelled to do so by some other body acting upon it. Neither does a non-living object move itself in any way. It will remain where it is, and as it is, for an indefinite length of time — as long as no other body acts upon it. These three facts have been embodied in what is called a law of motion, namely: *Every body will maintain its state of rest or of uniform motion in a straight line as long as it is not acted on by another body.* This action of bodies of all kinds and of all sizes is sometimes called the *Law of Inertia.* It means that in the absence of friction and impact, bodies in motion will continue to move in the same direction and at the same rate forever.

It is because the earth is moving in frictionless space that it continues to rotate so regularly upon its axis, so that the length of the day has not varied a second in 2000 years. Uniform motion under such conditions does not imply that any power or force or pressure is acting upon a body to keep it going. Wherever there is friction, a continuous push is needed to maintain uniform motion, and the push must be equal to the frictional resistance.

Acceleration. — When the push of another body exceeds the frictional or other resistance, then the body pushed has its velocity increased, and if the pressure be constant the increase in velocity is constant; that is, if a pressure of, say, five pounds will give a body a velocity of two feet in one second, it will give it an additional velocity of two feet the second second, and still another addition of two feet the third second, and so on. That is to say, the velocity will be proportional to the time the pressure is maintained. In such a case the *rate of increase in velocity is called acceleration*, which in the above example is two feet. If the pressure were greater the acceleration would be greater. A body with an acceleration of two feet per second will have a velocity of four feet at the end of the second second, and twenty feet at the end of the tenth second.

If we let v equal velocity, t, time, and a, acceleration, then $v = at$ and $a = \frac{v}{t}$.

Gravity produces acceleration upon falling bodies. It is found by experiment to give a velocity of 32.2 feet per second to any freely falling body; hence, if a body falls for two seconds its final velocity is 64.4 feet, and so on. The letter g is commonly employed to represent the acceleration of gravity, and in the above expression would take the place of a, which then would read $v = gt$. This will be the meaning of the letter g hereafter, and its numerical value will be 32.2 feet. Of course gravity retards the velocity of a body thrown up into the air, and its retarding pull subtracts from the velocity of an ascending body 32.2 feet for every second. This is called *negative acceleration.* It is found by experiment that a body falling from rest during one second falls but 16.1 feet while it is acquiring a final velocity of 32.2 feet. Its average velocity for the whole second is one half its final velocity. In like manner the average velocity of a falling body for any time is half its final velocity, or $\frac{gt}{2}$. Thus, a body falling from rest for five seconds will have a final velocity of $gt = 32.2 \times 5 = 161$ feet, and an average velocity of $\frac{gt}{2} = \frac{161}{2} = 80.5$ feet. If it falls for ten seconds, $gt = 322$ feet, and its average velocity is $\frac{gt}{2} = 161$ feet, and so on for any value of t.

The space, s, that a falling body will move through will be equal to its average velocity multiplied by the time, $\frac{gt}{2} \times t$.

Thus, if the body fall for five seconds its average velocity will be $\frac{gt}{2} = 80.5$ feet, and the space s will be $\frac{gt}{2} \times t = 402.5$ feet. If it fall for ten seconds its average velocity will be $\frac{gt}{2} = 161$, and the space s will be $\frac{gt}{2} \times t = 1610$ feet.

In this the prime factor t appears twice, and it is customary to write the expression thus, $s = \frac{gt^2}{2}$, and to say that the space passed over by a falling body is proportional to the square of the time. The same law holds true for any other constant pressure or acceleration in any direction, and a, acceleration, may be substituted for g in the formula; thus, $\frac{at}{2}$ = average velocity, and $\frac{at}{2} \times t = s$, space passed over. For example, a constant pressure gives an electric car an acceleration of 5 feet per second, what is its velocity in 8 seconds, and how far will it have moved?

Ans. $at = 5 \times 8 = 40$ feet per second, $s = \frac{at}{2} \times t = \frac{5 \times 8}{2} \times 8 = 160$ feet.

It is to be understood that in all practical problems of moving bodies in the air, the air itself acts to retard somewhat their movements. The value of g as given, 32.2 feet, represents the velocity that a body falling in a vacuum would acquire. Above the atmosphere there is nothing to retard motions, and meteors acquire a very high velocity (25 or more miles a second), and when they reach the atmosphere on their way to the earth, the friction is so great as to set them afire, which shows itself in the glow in their path. This high velocity is not due to the earth alone, but to the sun also. As the sun is much larger, its attractive power is greater, and therefore the value of g at the sun is about 28 times its value at the earth, that is, about 900 feet. A body drawn to the sun may have a velocity of 400 miles per second, the highest translational velocity known.

The curvature of the earth is 8 inches to the mile, that is, if a long straight horizontal line were to touch the smooth surface of a large body of water, like a lake or ocean, it would be eight inches away from the surface at the distance of a mile. If, then, above the atmosphere, one were to shoot a rifle ball in a horizontal direction with such velocity that it would go a mile while the body were falling eight inches, the ball would be no nearer the earth than when it started, and if not retarded by friction it would continue on, quite round the earth, — would indeed become a satellite. Can you compute what the velocity would have to be?

QUESTIONS.

1. If a body could be shot off in free space, beyond the atmosphere, what would be its direction, and what its change of velocity, if any?

2. If a body move with uniform velocity 144 feet in three seconds, what is its velocity per second?

3. A body has an acceleration of five feet, what will be its velocity at the end of five seconds? What will be the whole distance passed over? What will be its velocity at the end of the third second? What space will it pass over during the last second?

4. If a body be tossed up into the air with a velocity of 32.2 feet per second, how long a time will it continue to rise? How high will it rise? How long will it remain in the air?

5. How long will it take a body to fall 1000 feet, ignoring the resistance of the air?

6. If a bullet were shot vertically upwards with a velocity of 900 feet per second, how high would it rise, and how long would it be rising?

7. If the same thing should be done at the sun, how high would it rise, and for how long?

8. Find what the velocity of the moon is in its orbit at the distance of 240,000 miles.

CHAPTER III.

WORK.

A *push* or a *pull* may be measured in pounds. A spring balance answers both for weighing things — finding gravitative pressure — and for measuring pressure in any other direction. The amount of pull is indicated in pounds and fractions of a pound upon the scale. If the pressure[1] results in the movement of the body pulled, work has been done; if the body does not move, no work has been done. *The amount of work is equal to the pressure measured in pounds multiplied by the distance in feet, and the product is called foot-pounds.* Thus, suppose that to move a carriage it requires a pressure of 20 pounds. If that pressure be maintained and the carriage be moved ten feet, then $20 \times 10 = 200$ foot-pounds of *work* has been done. Unless the carriage moves there is pressure, but no work. This is the sense in which the term is used in physical science. If we let p stand for pressure in pounds, d stand for distance in feet, and W for work, then $pd = W$ in foot-pounds.

If a pound weight rests on the table, its pressure is constant there, but if the weight be raised a foot, the pressure will be constant through the distance of a foot,

[1] The word *pressure* is substituted for the word *force*, which is common in other books, thus gravitative pressure instead of the force of gravitation. There are many definitions and indefinite suggestions to the word force, while in physics it means what is meant by the word pressure, which word has no abstract implication.

and a foot-pound of work will be done. Hence 10 pounds raised 1 foot is the same in work as 1 pound raised 10 feet.

If a man weighing 150 pounds walks upstairs 10 feet he has done 1500 foot-pounds of work. If a hod-carrier carries 100 pounds of brick up a ladder 20 feet he has done $100 \times 20 = 2000$ foot-pounds, aside from that represented by his own weight raised the same distance; whether he goes fast or slow makes no difference in the amount of work done.

Power has to do with the *rate at which work is done*, and rate implies time. The unit of time is the second, and the amount of work to be done in a second has been set by the demands of business and is known as the *horse-power*. It equals 550 foot-pounds per second or 33,000 per minute. Hence,

$$\frac{\text{Pressure in pounds, } p, \times \text{ distance in feet per second, } d,}{550}$$

= horse-power. For example: Suppose a pair of horses, by pulling 200 pounds steadily, is able to move a car at the rate of five feet a second, how much power is spent?

$$\textit{Ans.}\quad \frac{200 \times 5}{550} = 1.8 \text{ horse-power.}$$

Whether the pulling was done by horses, steam-engine, or electric motor would make no difference in this, so both steam-engines and electric motors are rated in horse-power.

From the above principles it may be seen that no matter how large a body may be, nor how fast it may move, if it produces no pressure upon another body it does no work; also that friction, which is the resistance to motion of surfaces in contact, is proportional to pressure, as it may be measured in foot-pounds. A body like the moon may continue to go round the earth indefinitely long, though there is no pressure acting.

If a ten-pound weight rests upon the ground it does no work, yet if a man holds it up for a time it tires him, and he is conscious of having done work in the physical sense, although he has held it quietly. The explanation is that when muscles are stretched they begin to vibrate longitudinally, a great many times a second. As they lengthen the weight falls slightly, as they contract it is pulled up, and thus the effect is the same as though one had raised and lowered the weight successively at a slower rate. As one walks his body rises and falls regularly; his muscles therefore do work in raising his body an inch or two each step, and after a time he becomes weary. Suppose he raises his body of 150 pounds two inches each step, and that he steps twice a second, he is doing $\frac{150}{6} = 25$ foot-pounds of work each step, 50 foot-pounds each second, equal to $\frac{50}{550} = \frac{1}{11}$ of a horse-power. In flight, birds propel themselves through the air by beating it with their wings, and thereby producing pressure and doing work. If a large bird like a goose were flying at the rate of thirty miles an hour, which is equal to 44 feet a second, and if a continual pressure of one pound were maintaining that speed, it would be expending 44 foot-pounds of work per second, nearly one-twelfth of a horse-power. It is not probable that the pressure would be nearly so much; and if it were but one ounce, the goose would be expending $\frac{44}{16} = 2.75$ foot-pounds per second, the two-hundredth of a horse-power.

QUESTIONS.

1. How much work is done in pushing a cart a mile if the pressure be 10 pounds?

2. How much work do you individually do in climbing a hill 500 feet high?

3. If a squirrel carries five pounds of nuts to his storehouse fifty feet up a tree, how much work does he do?

4. A wagon is loaded with stones that have to be raised five feet; if the load weighs 1000 pounds how much work was expended in loading it?

5. The pull of a locomotive upon a train going 30 miles an hour is 2000 pounds, what horse-power is expended?

6. The balance-wheel to an engine doing 40 horse-power of work is revolving four times a second and is ten feet in diameter, what is the pull upon the belt?

7. If a pound weight be tossed up 16 feet high, how much work is represented? What initial velocity is needed to make it rise 16 feet?

8. If a bullet weighing an ounce be shot vertically upwards with an initial velocity of 1000 feet per second, how high will it rise, and how much work will be spent on it?

9. If a laborer shovels six pounds of dirt each shovelful, and ten of them per minute, into a cart six feet high, at what rate is he doing work?

10. A pressure of two pounds is needed upon the crank of a grindstone that has a radius of one foot. If it be turned twice a second, how much power will be spent?

11. If atoms be the fifty-millionth of an inch in diameter, at what rate per second are they assembled and arranged in a stalk of asparagus, half an inch in diameter, that grows six inches in length in a day?

CHAPTER IV.

ENERGY.

BY energy is meant *ability to produce pressure.* Wherever there is pressure of any kind there is energy, and where the pressure produces motion, it does work in the mechanical sense — a pressure multiplied by a distance. The pressure of the wind upon the sails of a vessel or of a windmill causes the one to move forward and the other to rotate. The pressure of water causes the water-wheel to turn, the pressure of steam causes the piston to move forward, the pressure of an electric current causes the armature of an electric motor to rotate, and the amount of work is proportional to the pressure.

When a body is subject to two equal and opposite pressures it cannot move, of course. A paper-weight lying on the table presses upon the table, and the table presses upward upon the weight equally; and this condition of two equal and opposite pressures on a body, wherever it occurs, is called *equilibrium*, and no motion results. If the weight were free to fall, that is, if the table pressure were removed, the gravity pressure would still remain, and it then could do an amount of work proportional to the distance it could fall. If it weighed one pound and could fall three feet it would do three foot-pounds of work; if the weight could fall 25 feet it could do 25 foot-pounds, and so on.

There are many cases of equilibrium where there is pressure, but no motion is apparent. A clock wound up with pendulum stopped, a steam boiler with valves closed, water in any containing vessel, a magnet with armature held near it, are common examples. The energy in these cases is called *potential energy.* When a given pressure upon a body is not balanced by an opposite one, the body moves, and may do work. The energy of a moving body is called *kinetic energy.*

Rate of Work. — Wherever work is being done it is done at some *rate* which implies time, and it is often convenient to know what the working power of a moving body is when we know only its weight and velocity.

We have seen that if a body be tossed up into the air with a velocity of 32 feet per second, it will rise 16 feet; if the body weighed a pound then the work done on it would be 16 foot-pounds, for it represents a weight of one pound raised 16 feet. If the initial velocity were doubled and made 64 feet per second, then the body would rise to the height of 64 feet. 64 foot-pounds of work would be done by doubling the velocity, and 64 is four times 16; that is, by doubling the velocity the amount of work has been increased four times, hence we say that the work a body will do is proportional to the square of its velocity. Evidently a pound falling 16 feet will do 16 foot-pounds of work, for it would raise a pound 16 feet high, and in falling 64 feet it will do 64 foot-pounds. Now, the distance a body will fall in any time is equal to the square of its final velocity divided by twice the value of gravity, g.

$$\text{Distance, } d = \frac{v^2}{2g}.$$

On page 21 it is shown that the weight of a body is a variable quantity, and consequently its pressure varies with its place on the earth. If the weight of a body be divided by the value of g wherever it may chance to be, we have an invariable quantity, which is called the mass of the body.

$$\frac{w}{g} = m, \text{ and } w = mg;$$

that is, the weight of a body is equal to its mass, m, multiplied by the value of g.

The product of the weight of a body, which represents its pressure, into its distance moved against gravity, $pd, = \text{work} = W;\ p = w = mg;\ d = \frac{v^2}{2g}.$

Using these values of p and d we have

$$mg \times \frac{v^2}{2g} = \frac{mv^2}{2} = \textit{work in foot-pounds};$$

that is, the work is equal to one-half the product of the *mass* into the square of the velocity, and this quantity represents the energy which a mass m has when its velocity is v. One must keep in mind that m is not in pounds, for a pound weight has 32.2 units of mass hence a unit mass is $\frac{1}{32.2}$ of a pound or about half an ounce. If we now substitute for m its equivalent $\frac{w}{g}$ in the expression, we have $\frac{wv^2}{2g}$ = energy, and one may compute how much work any body may do that has a known weight and velocity. For example:

A body weighing 100 pounds has a velocity of 25 feet per second, how much energy has it?

Ans. $\frac{wv^2}{2g} = \frac{100 \times 25^2}{2 \times 32} = \frac{62500}{64} = 976$ foot-pounds.

Vibratory Energy. — So far energy of translatory motions has been considered; but if a body vibrates like a tuning-fork it has energy, for if a light body be made to touch one of the prongs it will be beaten away.

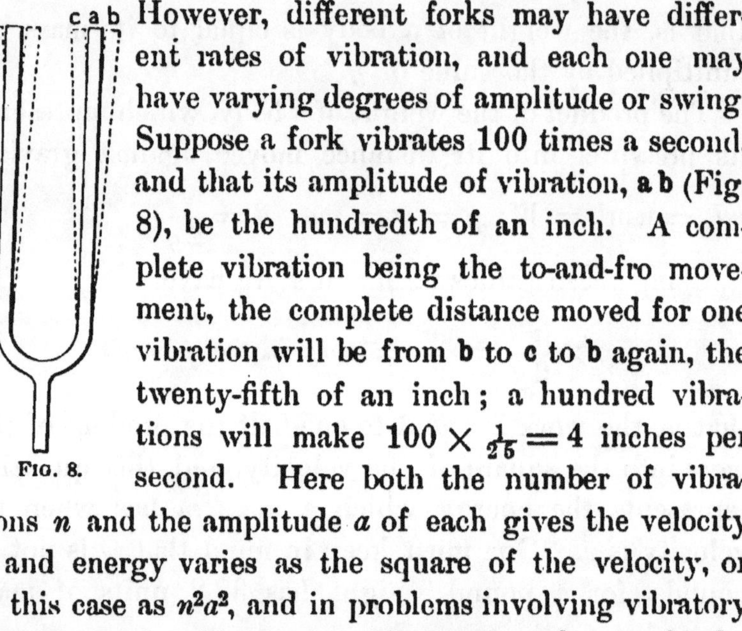

FIG. 8.

However, different forks may have different rates of vibration, and each one may have varying degrees of amplitude or swing. Suppose a fork vibrates 100 times a second, and that its amplitude of vibration, **a b** (Fig. 8), be the hundredth of an inch. A complete vibration being the to-and-fro movement, the complete distance moved for one vibration will be from **b** to **c** to **b** again, the twenty-fifth of an inch; a hundred vibrations will make $100 \times \frac{1}{25} = 4$ inches per second. Here both the number of vibrations n and the amplitude a of each gives the velocity v, and energy varies as the square of the velocity, or in this case as n^2a^2, and in problems involving vibratory movements it will not be sufficient to take simply the number of vibrations for the velocity, so vibratory energy will be represented by $\frac{mn^2a^2}{2}$ or $\frac{wn^2a^2}{2g}$.

Rotary Energy. — If an iron rod a foot long and an inch thick were to be given a translatory velocity lengthwise of, say, 10 feet a second, it would have a definite amount of energy. Suppose the rod were curved into a ring, and made to spin around an axis in the middle of the ring at such a rate that a point on it moved 10 feet a second, its energy would be the same as before, because there was the same mass moving with the same velocity; but it would not be advancing, because its motion was rotary. Given the number of rotations a body may make per second, its weight and diameter if it be a wheel or a disk, one may compute the corresponding velocity as if it were translatory, and determine the energy. Thus, a wheel ten inches in diameter weighs 25 pounds, and has most of its weight in its rim. If it revolves ten times a second, how much energy has it?

Ans. $w = 25,$

$$v = 10 \times 3.14 \times 10 = \frac{314}{12} = 26.1 \text{ ft. per sec.}$$

$$\frac{wv^2}{2g} = \frac{25 \times 26.1^2}{64} = 266 \text{ foot-pounds.}$$

QUESTIONS.

1. If a body weighing 10 pounds be shot vertically upward with a velocity of 96 feet per second, how much work has been done on it? How much work will it do when it falls?

2. If you were to catch a brick weighing 6 pounds that had been tossed up 10 feet to you, how much energy would the brick have? Would you be conscious of expending energy in holding it? What kind of energy would you say the brick had? Were you to drop it, what amount of energy would it have 5 feet

below? If it were to be stopped there and again dropped, would it make any difference in the whole amount of energy the brick had or the amount of work it could do?

3. If a ton of water falls 10 feet, how much work is it capable of doing?

4. What velocity does water acquire in falling 20 feet?

5. A certain stream has 5 tons of water going each minute over a fall 30 feet high; what horse-power does it represent?

6. A meteor weighing 1 pound comes into the atmosphere at the rate of 25 miles per second; how much energy has it? If it was applied to lift a ton weight, how high would it raise it?

7. A cannon ball has a velocity of 1000 feet per second. If it weighs 500 pounds, how much energy has it? How long would it maintain a horse-power if it could be applied?

8. If a person weighing 150 pounds runs upstairs at the rate of 4 feet per second, what horse-power does the work represent?

9. If a person weighing 160 pounds jumps 6 feet high, how much work does he do? If he reaches that height in a second, what horse-power does he expend?

10. The rim of a balance-wheel is 10 feet in diameter and weighs 5 tons. If it be driven around 4 times a second, how much energy will it have?

(NOTE. — The use of a balance-wheel is to supply energy for a short time when for any reason the engine fails to keep up the speed.)

11. A pile-driver weighing 2500 pounds falls 15 feet and strikes the head of a pile; what velocity does it have, and how much energy?

12. A 10-horse-power engine can lift 15 tons to what height in 30 seconds?

13. A loaded wagon weighing a ton is drawn up a hill a mile long, the top of the hill is 2000 feet higher than its base. How much work is done? Would it make any difference whether it is drawn by horses or by a steam-engine? whether it were done quick or slow?

CHAPTER V.

MACHINES.

WHENEVER it is desirable to transfer pressure from one place to another, we employ a device which we call a *machine*. When a weight like a stone is raised with a crowbar, the pressure that is applied at one end is transferred to the other and utilized there. The bar acts to transfer and change the direction of the pressure by means of the support provided called a *fulcrum* **a**;

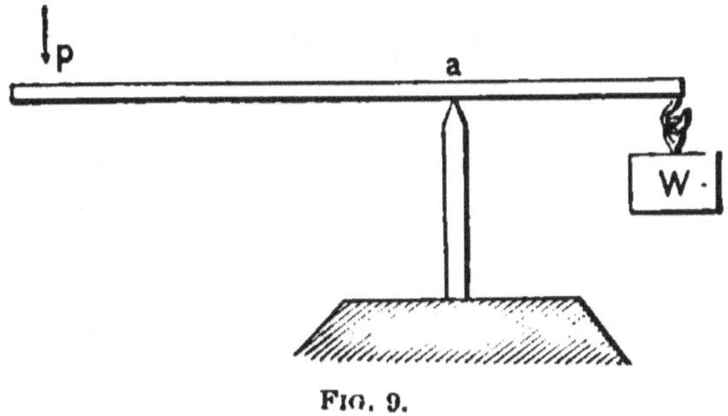

FIG. 9.

for when the pressure is downward at **p** it is upward at **w**. Such a machine is called a *lever*, and the principle of *work* applies to it. Thus, if at **p** there be a downward pressure of 10 pounds continued through one foot, then 10 foot-pounds of work have been done at the other end of the lever, whether it be long or short. In any case, the product of the weight (pressure) at **w** into the distance it is raised will be equal to the foot-pounds of work spent at **p**. The actual distance **w** will rise depends upon the relative lengths of the parts **a p** and **a w.** If **a p** be twice as long as **a w**, the distance **p** will move will be twice the distance that **w** will move.

Suppose a lever **p w**, four feet long, has a fulcrum at **a** one foot from **w**, and that at **w** is a weight of 50 pounds, what pressure at **p** will be required to balance **w**?

Ans. Length $\mathbf{aw} \times \mathbf{w} =$ length $\mathbf{ap} \times \mathbf{p}$; length $\mathbf{aw} = 1$; length $\mathbf{ap} = 3$; $1 \times 50 = 3 \times \mathbf{p}$; $\mathbf{p} = \frac{50}{3} = 16.6\dot{6}$ pounds.

Again; if pressure at **p** be continued through a foot, to what distance will **w** be raised?

Ans. Work at $\mathbf{p} =$ work at $\mathbf{w}$; that is, pd is the same at both places; hence $16.66 \times 1 = 50 \times d$. $d = \dfrac{16.66}{50}$ of a foot.

The product of the pressure into the distance through which it is maintained, will give the work done at the other end of the lever. Any amount of work done at one end of a lever will do an equal amount of work at the other end. Examples of the lever are very common, — the claw-hammer, pincers, scissors, the weighing balance, and the steelyard. These and others may be thought out with reference to the above principles.

The *pulley* is a machine for changing the direction of a pressure with a rope or belt. In its simplest form it consists of a grooved wheel around which a rope may be passed (Fig. 10). It is used for conveniently lifting bodies to considerable heights by a continuous pull or pressure at one end of the rope; the weight **a** will be balanced by an equal weight or pull upon the other end of the rope **b**, and if the pressure upon **b** be downward, of course it does work equal to *pd* upon the weight **a**. If two pulleys be used, and one end of the rope be made fast, as at **d** (Fig. 11), then half of

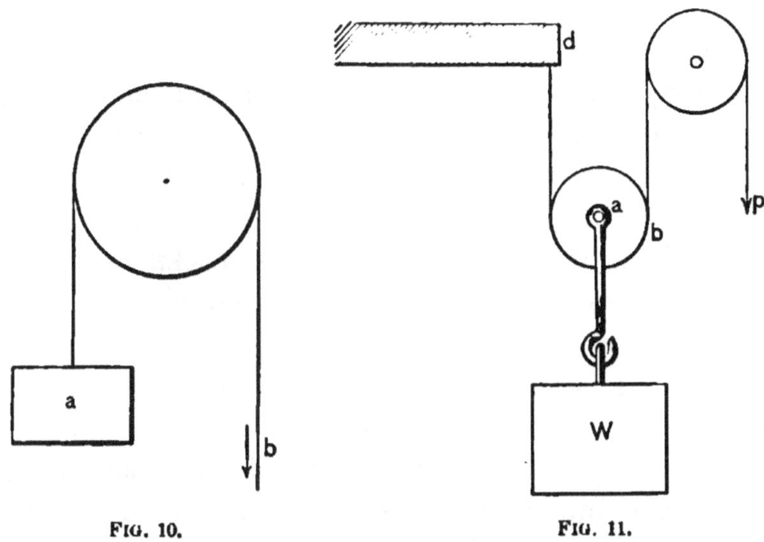

Fig. 10. Fig. 11.

the weight of **w** is supported by the beam, and to raise the weight **w** one foot the rope at **p** must be pulled through two feet; so by pulling 50 pounds two feet it becomes possible to raise 100 pounds one foot.

Can you devise a pulley arrangement by which a man may raise himself to any height?

Transference of Pressure and Power.—In the pulley we have the translatory motion of the rope in the direction of the pressure, changed into rotary motion. In the wheelbarrow and in common carriages there is translatory motion changed into rotary, and in the locomotive rotary motion of the drivers changed to translatory of the whole engine, also oscillatory of the piston into rotary and various others. In the sewing-machine it is plain how one kind of motion is changed into another kind with the corresponding pressures. The various levers, pulleys, and other parts, act to transfer the pressure from the feet where it begins on the machine to the active needle and shuttle, and at the same time to give to them their appropriate motions.

The combination of pulley and belt is the chief mechanism in machine shops, and factories, for the distribution of power to the various machines. They deliver pressure or mechanical power, and no more than is given to them. They add nothing to energy, but always subtract somewhat from it because of their necessary friction. In a machine shop the power to run the machines when none of them are doing their appropriate work is generally much greater than that required for their work. The amount of work a belt can carry depends upon its velocity. Suppose a belt when running maintains a constant pull of 100 pounds. If it moves at the rate of 50 feet per second it transmits, $pd = 100 \times 50 = 5000$ foot-pounds in that time; if it runs 25 feet per second it transmits but one-half of that.

The furnace of a steam-engine is a machine for transferring the energy of fuel to the boiler; the boiler and pipes transfer it to the engine, the engine to the shafting by belts, and the pulleys on the shafting to the individual lathes, planers, looms, or whatever kind they may chance to be. From beginning to end it is a pressure, energy, the power to do work in one form or another, that is modified in direction or form or rate by the various devices by which it is distributed, and which are called machines.

Transformation of Motion.—When one kind of motion is changed into another, as translatory into rotary, the resultant is called *transformed motion*, and some kind of machine is always needed to effect the transformation. The quick unwinding of the string in translatory motion gives the top its spin. The rotating wheels of a locomotive give to it its forward translatory motion. The oscillations of the piston give rotation to the wheels, and the continuous pressure of the steam, due also to its molecular motions, is transformed into the oscillatory motion and pressure of the piston in the cylinder. Another study of the sewing-machine will give one a good idea of the mechanism by which one kind of motion is transformed into another; and it is a capital exercise to invent ways for thus transforming one kind of motion into another of a given kind. All the various things done by machines are brought about by changes in the character of the motion of the various parts. The weaving of carpets and of cloths with intricate patterns goes on in a loom in an automatic way that is wholly mechanical, though it looks as if there were intelligence directing the various parts. There is no movement requiring skill of the hands but can be duplicated by machinery; even the piano-touch of a skillful musician is but a mechanical movement varying in muscular quickness, pressure, and release at proper times, and can be done by a machine having a proper degree of perfection. If a mechanically played piano is not satisfactory music now, it is because the mechanism is not perfected. Study the action of a piano-key and see if this conclusion is not warranted. Do you think that with a motion of your finger you can produce any motion that mechanism could not?

These mechanical principles are not restricted to masses of matter of definite or visible size, but are as applicable to microscopic particles, molecules, and atoms as well, so that exchanges of energy among them take place whenever there is a change in the character of the motions. For convenience, we speak of the motions of masses of visible magnitude as *mechanical motions*, to distinguish them from such as are too minute to be seen, which are called *molecular* or *atomic* motions. Thus the motions of an engine, of a grindstone when turned, of a swinging pendulum, or of the balance-wheel of a watch, are called *mechanical*, and the energy they represent is *mechanical energy;* while the motions in a gas, whether of one kind or another, are *molecular*, and the energy is *molecular energy*. Yet it should be borne in mind that this naming is only for convenience, and does not imply or represent any fundamental difference in matter of different sizes.

QUESTIONS.

1. A man weighing 150 pounds on the end of a crowbar 5 feet from the fulcrum is just sufficient to raise a stone on the short end of the bar 6 inches from the fulcrum; what is the weight of the stone?

2. In what ways can a bar, pole, or rod be used to do work?

3. Can a rope be substituted for any of these?

4. What is the difference between a push and a pull?

5. A belt is transmitting 10 horse-power to a pulley 5 feet in diameter; what is the velocity of the pulley?

6. If an air-molecule be moving in its free path 1500 feet per second, and its free path be the $\frac{1}{250000}$ of an inch, how many times does it strike other molecules and change its direction in one second?

7. The crank of a grindstone is one foot long. If it requires a constant pressure of one pound to rotate the stone, how many foot-pounds of work are done in one revolution?

8. If the grindstone be 3 feet in diameter, what pressure upon its face will be needed to hold it still if the push upon the crank be 10 pounds?

CHAPTER VI.

GASEOUS PHENOMENA.

The Air. — The atmosphere that surrounds the earth is a body of gas which is held to the earth by gravity in the same way that other bodies are held. It turns round with the earth, as a part of it. At the equator the velocity of the earth is more than a thousand miles an hour; at the poles it is nothing. Between the two places there is a constant exchange, which gives rise to extensive air-currents we call winds. Because the air has mass, the wind results in pressure. The velocity of the wind is measured by a machine called an anemometer, a kind of small windmill, and its pressure varies as the square of its velocity. Wind-velocity varies within wide limits at the surface of the earth, from no wind to tornado velocities, which are 150 or more feet per second. This wind-pressure is utilized by windmills for purposes where a slight amount of power, such as pumping water, is required. Wind-pressure at moderate velocities is not very great, being for one mile an hour only .005 of a pound per square foot; but as its pressure increases as the square of the velocity, at 10 miles per hour it will be .5 of a pound, and 50 pounds at 100 miles per hour. A wind-wheel 25 feet in diameter, in a wind blowing 15 miles an hour, will give but about one horse-power.

Wind-pressure is the result of a current or a body of air moving in some given direction, and is thus to be distinguished from another kind of pressure in the air, which is due to the individual motions of its molecules. The latter pressure depends upon density and temperature, and exists whether the wind blows or not. This is called *gaseous pressure.*

The density of the air means the number of molecules per cubic inch; the temperature depends upon the rate of vibration of the molecules, which gives them their translatory motion. There are so many molecules in a cubic inch that each one can move but a short distance before it bumps against another, the impact changing the direction of both. The distance each one moves between collisions is called its *free path.* The distances are not uniform, but the average of the distances is called the *average free path* of the molecule, which, of course, depends altogether upon the density of the gas. Each molecule of air or of any gas in a vessel is free to move in any direction, and is moving at the rate of about a quarter of a mile per second; but its path is short. Some, at the inner surface of the vessel, bump upon that surface and press outwards. Many millions per second do this, and the result of such a bombardment on the inner surface is a continuous outward pressure on each of the walls. This amounts ordinarily to nearly 15 pounds per square inch of surface. Of course there is the same kind of action and pressure outside as inside any vessel, and so there is equilibrium of pressure, as illustrated on page 43 by the weight on the table. If we contrive to extract the air from any inclosed space we have what is called a vacuum; but the external pressure still remains. A vacuum may be produced by a machine called an *air-pump*.

It consists of a cylinder **B**, provided with a movable piston **P** which has a valve **s** in it opening upwards. At the bottom of the cylinder is another valve **t** also opening upwards. A tube **T** leads from the cylinder to a flat plate **L**, upon which glass jars **R**, called receivers, may be placed. These receivers have carefully ground edges so as to be air tight when in place.

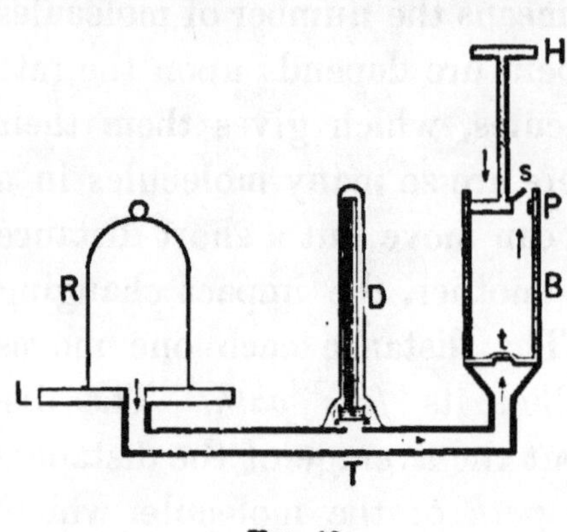

FIG. 12.

A barometer **D** is sometimes connected to the tube **T** to indicate the degree of vacuum produced. When the handle **H** is pushed down, the air below the piston is compressed, valve **t** being closed. The increased pressure opens **s** and allows the air to escape. When the handle is pulled up, the valve **s** is closed and the air in the receiver and tube opens the valve at **t**, as the gaseous pressure is greater below the valve than above it. Each upward pull removes some of the air from the receiver into the cylinder and each downward push permits the escape of most of the air above the valve **t**. With an ordinary pump of this kind one may extract 99 per cent of the air from a receiver, and consequently leave an internal pressure in it of only $\frac{15}{100}$ of a pound per square inch. With such a machine a great variety of experiments in air-pressure can be made.

On page 7 it is shown what an enormous number of molecules there is in a cubic inch of matter at ordinary pressure. If the gas-molecules be reduced to one-hundredth of their number, one can see from the figures that the number of molecules will not be apparently much changed,—21,000000,000000,000000. Divide it by a million and there is still an astonishing number left. Dividing by a million will reduce the pressure a million times; but it is difficult even to approach a perfect vacuum. By the most perfect means at our disposal now, it is possible to reduce the pressure to the hundred-millionth part of 15 pounds per square inch.

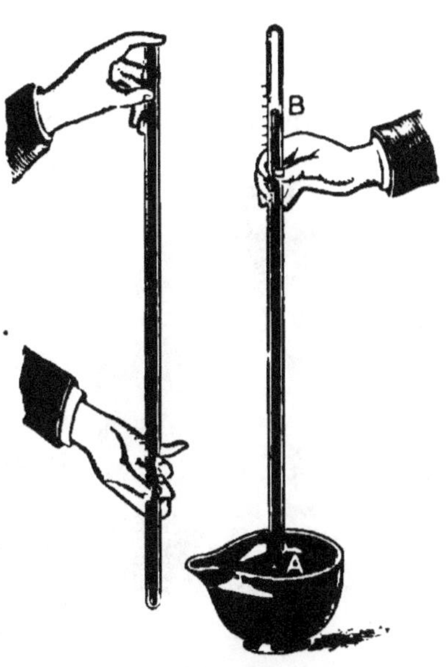

FIG. 13.

The pressure of the air may be measured by providing a tube 32 or 33 inches long, filling it with mercury, and then carefully inverting it in a cup of mercury (Fig. 13), not allowing any air to enter the tube. The air-pressure will then be able to balance the pressure of a column of mercury, which is usually about 30 inches high, but varies from that an inch or more either way. Such a machine is called a *mercury barometer*, and it has important uses in the study of the weather. We may find what the pressure of the air is by weighing the mercury in the tube. If the column of mercury be a square inch in section, and 30 inches high, which just balances the pressure of the air, it will equal the weight of 30 cubic inches of mercury, which is nearly 15 pounds.

Another kind of barometer, called the *aneroid* (Fig. 14), is in common use. It consists of a shallow metallic box, having a thick back and a thin, flexible face. This box is partially exhausted of air and then hermetically sealed. The thin face moves in and out as the air-pressure is greater or less, and this movement of the face acts upon a mechanism for changing the direction of motion, and for amplifying it by a hand that moves over a circular scale, where its movements may be observed.

FIG. 14.

When air is condensed by forcing it to occupy less than its usual volume, it is found that its pressure *increases as its density;* so that if a cubic foot at 15 pounds per square inch pressure be reduced to half a cubic foot, the pressure will be 30 pounds to the square inch. If it be made less dense by giving it double the space, its pressure will be reduced to 7½ pounds, and so on. Thus it appears that *the density of a gas varies inversely as the volume it is made to occupy.* This is known as *Boyle's Law.* That its pressure should vary as its density follows from the fact that gaseous pressure upon a surface depends upon the number of molecules that strike the surface at a given instant, and is proportional to that number.

The higher one climbs a mountain, the less the number of molecules in the air per cubic inch, and consequently the less the pressure; and this is shown by the barometer. If one observe the barometer to stand at the height of 30 inches at the sea-level, and then carries it up on a hill, mountain, or in a balloon, he will see that at the height of 900 feet the mercury will have fallen about an inch; at the height of a mile, 6 inches, at the height of 3 miles, nearly 15 inches. Almost half of the atmosphere is below the latter elevation. With a good barometer one may easily observe the difference in pressure between the ground and the second story of a house.

Gaseous Buoyancy. — A hollow metallic globe (Fig. 15) that, whether full or empty, keeps the same volume, weighs less when the air has been removed from it by an amount equal to the weight of the air displaced. A soap-bubble blown with common gas will rise rapidly in the air, for it weighs less than the air that it displaces. Suppose the bubble to have a volume of 100 cubic inches, — it will displace 31 grains of air. If the weight of the bubble and its contained air be 20 grains it will be subject to an upward pressure of 31 — 20 = 11

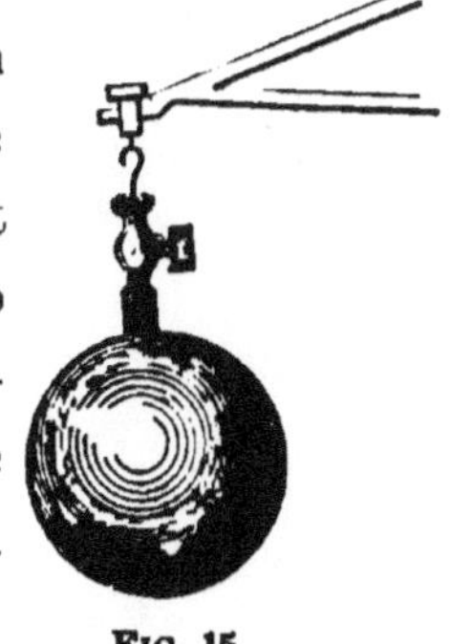

Fig. 15.

grains; that is, the difference between its total weight and the weight of the air displaced by it. This is the principle of the balloon. Let a balloon and its contents weigh 100 pounds, and its volume be great enough to displace 300 pounds of air, — it will have an ascensional pressure of 200 pounds, and can easily carry up a man of ordinary weight.

CHAPTER VII.

LIQUID PRESSURE.

THE same principles that apply to gases apply equally well to liquids. The weight of a cubic foot of water is 62.5 pounds; consequently this is the weight that must be supported in order to sustain a cube of water one foot square. If another cube be put on top of the first, the pressure on the bottom will be $62\frac{1}{2} \times 2 = 125$ pounds. The pressure is proportional to the depth. Hence the pressure per square foot for any depth is equal to $62\frac{1}{2}$ multiplied by the depth in feet. The pressure on a square inch is the $\frac{1}{144}$ of what it is for a square foot: $\frac{62.5}{144} = .435$ of a pound. If a tube 35 or 40 feet long be filled with water, as the barometer tube was with mercury, and then be inverted in a tub of water, the water will stand at such a height in it as will give a pressure of 15 pounds per square inch, the pressure of the atmosphere. If the tube be one square inch in section, then the pressure of one foot in length will be .435 of a pound, and to weigh 15 pounds the column must be $\frac{15}{.435} = 34$ feet long; that is, a column of water 34 feet long balances the pressure of the atmosphere on a square inch.

The common suction-pump (Fig. 16) utilizes this pressure, and with it water cannot be made to rise higher than 34 feet, and generally not so high, owing to imperfections in the joints which let air in between the valve and the top of the water in the pipe. The higher the hill or mountain on which the pump is placed, the less the distance to which air-pressure will raise water. The denser the liquid the shorter the distance it may be raised in this way. If some mercury be poured into a glass U-tube (Fig. 17), the surface of the mercury will be at the same height in both branches. If water be poured into one of the branches, it will so press upon the mercury as to lower the height in that branch and raise it correspondingly in the other. The column of mercury **ab** will now balance the column of water **dc**, and by measuring the length of these columns one can determine how many times one of these is longer than the other, which will give their difference in density. The column of water will be 13.6 times longer than that of the mercury column. The density of mercury is 13.6. If alcohol were used instead of water its column would be longer, for alcohol is less dense.

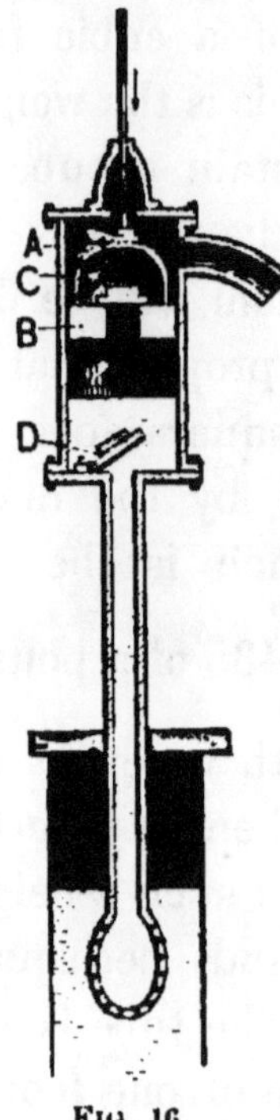

FIG. 16.

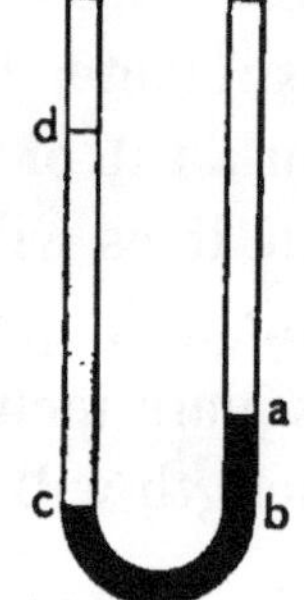

FIG. 17.

Specific Gravity. — When a cup is placed upon the surface of water it floats. It has displaced some of the water, enough to balance the weight of the cup. If the cup be pressed still deeper it will displace still more water, and the pressure upwards will be proportionally greater. Whatever be the weight or solidity of a body which is submerged in water it will displace a volume of water equal to its own volume, and will be pressed upwards so as apparently to lose as much weight as is equal to that of the displaced water. Thus, if a body displaces a pound of water it weighs a pound less in water than in the air. If a body weighs seven pounds in the air, and only six in the water, the loss in weight is one pound. The ratio of the weight of a body in air and its loss in water is called the specific gravity of the body.

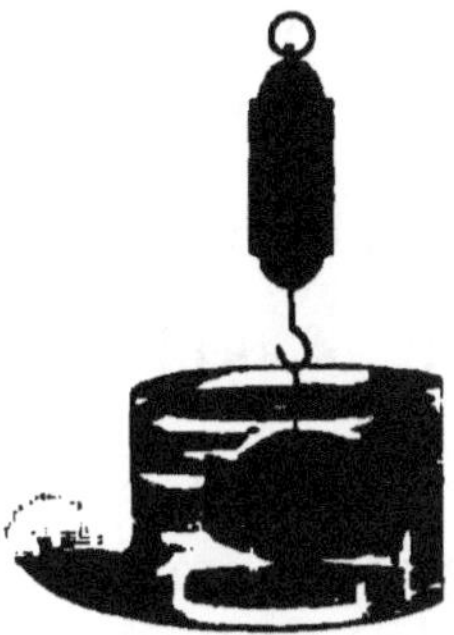

FIG. 18.

Let w be the weight of a body in air and l the loss when weighed in water (Fig. 18), then $\frac{w}{l}$ = specific gravity. There is a distinction between the specific density of a body and its specific gravity, for specific density

$$= \frac{\text{density of the body}}{\text{density of the water}},$$

and specific gravity $= \dfrac{\text{weight of the body}}{\text{weight of an equal bulk of water}}$.

But the two ratios are numerically equal, so that the number for the specific gravity of a body is the same as for its density.

The specific gravity of a liquid is its weight compared with the weight of an equal volume of water.

The specific gravity of a liquid may be determined

by carefully weighing a vial, filling it with pure water, and weighing it again to find the weight of water it will contain. Then fill it with the liquid to be determined, and find its weight. Dividing the weight of the latter by the weight of the water will give the specific gravity of the liquid. Thus, if the vial weighs 500 grains, filled with water, 850, and filled with alcohol, 780 grains, the water weighs $850 - 500 = 350$ grains, alcohol $780 - 500 = 280$ grains, and $\frac{280}{350} = .8 =$ specific gravity of alcohol.

Flotation. — A piece of wood will float on water, and to make it sink an additional weight must be applied. A piece of solid iron will readily sink in water, yet a tin cup, which is chiefly made of iron, will float, and the largest steamships, which are made of iron, will float safely across the ocean. This can be understood by recalling that when a body is placed in

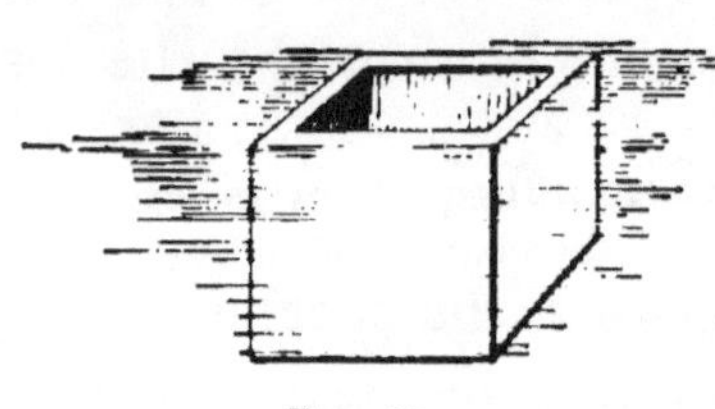

FIG. 19.

water it displaces some of the water and is pressed upwards by a pressure equal to the weight displaced. A cubic foot of iron will displace a cubic foot of water, and will lose $62\frac{1}{2}$ pounds in weight when immersed; but if the iron be shaped into a vessel so as to displace a weight of water equal to its own weight, it will just float (Fig. 19). The specific gravity of iron is 7.8, and if it can be so shaped as to displace 7.8 cubic feet of water it will float; and if, by making it thinner it can be made still more capacious, it can be loaded until the combined weights displace an equal weight of water. The weight of a vessel with its cargo is always equal to the weight of water which it displaces, or 62½ times the number of cubic feet of water it displaces. A cubic foot of gold in order to float would have to be so shaped as to displace 19 cubic feet of water, and a cubic foot of aluminum 2.6 cubic feet.

QUESTIONS.

1. If the pressure of the wind vary as the square of its velocity, what will be its pressure at 50 miles an hour, if at 1 mile per hour it be .005 pound?

2. At the top of Mt. Washington the wind has been observed to reach the velocity of 150 miles per hour, what then was the pressure per square foot?

3. What is the pressure upon the sails of a windmill with 100 square feet of sail surface when the wind blows 10 miles an hour?

4. What is the pressure per square inch in water at the depth of 5 feet?

5. If a cubic foot of air be taken when the pressure is 15 pounds per square inch, and it be immersed in water to the depth of 10 feet, how much will its bulk be reduced?

6. A brick is 8 inches long, 4 inches wide, and 2 inches thick; what weight of water will it displace?

7. A cubic foot of iron is so shaped that it just floats when placed upon water; what volume of water does it displace?

8. How much less will a cubic foot of iron weigh in water than in air? How much less will a cubic foot of gold weigh under similar conditions?

9. If a cubic foot of marble weigh 162 pounds, what will be its specific gravity?

10. If the specific gravity of silver be 11, what will be the weight of one cubic foot of it?

11. What will a gallon (231 cubic inches) of coal oil weigh if its specific gravity be .9?

CHAPTER VIII.

ON HEAT.

Let the hand be placed in contact with any object, as a book, a pencil, or a table. By the sense of touch we become conscious of the contact, and by moving the fingers upon it the same sense of touch informs us whether the surface is plane, curved, or angular, smooth or rough, so that even with the eyes closed we may determine the form of a body if it can be touched. If a piece of ice be touched, in addition to the sensation of contact of form and the character of the surface, a different sensation is perceived, which we call coldness. On the other hand, if a poker that has been a little while in glowing coals be touched, in addition to the sensation of contact there will be another sensation, which we call hotness, of which there are all degrees, so that we speak of hot bodies and cold bodies, and their difference in this respect we call their difference in temperature. Such terms as *warm* and *cool* are generally employed to denote temperatures that are agreeable. Such expressions as "a warm day," "a cool breeze" imply temperatures that are not unpleasant, while "a hot day" and "a cold wind" imply discomfort. In this way by our feelings we judge temperature. The sense of temperature, which all possess, does not tell us of the amount of the difference in temperature. If one should ask how much hotter a hot day is than a warm or a cool one, only some vague and indefinite answer could be given. We need to depend upon other senses than feeling to answer. Physiologists tell us that we have a particular set of nerves in the body, the function of which is to perceive heat, as we have nerves for the perception of touch, taste, and sight. A particular set of nerves implies a particular agency capable of acting upon them, and the agency for this set of nerves is called heat, but the phenomena of heat are apparently so unlike the phenomena of such mechanical bodies as we can see, and which have been considered in the previous chapters, that they form a separate part of physics, and introduce laws that will be new to the student.

THE ORIGIN OF HEAT.

I. Friction.—If one will rub his knuckles briskly upon his coat sleeve, he will find presently the heat sensation unbearable. A metalic button, rubbed in the same way on the floor gets too hot to hold with comfort in the fingers. Such mechanical action as this we call friction, and it has been found that wherever there is friction heat is generated. The friction brake on cars to bring them quickly to rest is a good example of this, for the brake when applied may be seen in the night to be giving a shower of sparks, and if touched after the cars have stopped may be found hot. In the absence of a supply of lubricating oil the friction on the car axles sometimes sets afire the waste in the boxes. Count Rumford found that water could be made to boil by the heat generated by boring a cannon; and Sir

Humphry Davy was able to melt two pieces of ice by rubbing them together. The scratching of a common match heats the end till it takes fire.

II. Impact. — A blacksmith may hammer a small piece of iron until it is too hot to hold. A bullet that has just struck the target is hot, and in the dark may be seen to produce a flash at the instant of impact.

III. Chemism. — When coal, wood, or any combustible thing is burned, there is much heat produced. This is a chemical phenomenon, and is due to chemical combination going on at a rapid rate. Also, if a test tube be filled to the depth of an inch with water, and an equal volume of strong sulphuric acid be then poured into it, the mixture will become too hot to hold. There is chemical action here, and if care be taken to observe accurately the volumes of the two liquids used, it will be found when the mixture has cooled that the resulting volume will not be equal to the sum of the volumes used.

IV. Electricity. — An electric current may produce a very high degree of heat. The electric light itself is due to the hot carbons, which shine.

In each of these cases it is to be observed that what immediately preceeds the appearance of heat is some kind of energy that is spent in producing the heat.

In the case of friction, the train of cars has a translatory mechanical motion, which the friction of the brakes stops, and in place of the mechanical motion heat appears. When the stroke of the hammer falls upon the nail, or the bullet strikes the target, there is the same translatory motion of a relatively large mass of matter suddenly stopped, the heat appearing as its substitute.

In like manner when air is suddenly compressed, as in the condensing syringe (Fig. 20), the heat resulting is so great as to ignite a piece of punk fixed at the end of the piston. This means that the relatively long free

FIG. 20.

paths of the molecules of the air are suddenly reduced, the molecules strike each other more frequently, and a flash may be seen if it be done in the dark.

When coal is burned its atoms combine with oxygen of the air, and there are violent atomic collisions producing a continuous flash, which we call a flame or glow; and when an incandescent lamp filament is made red-hot by the electric current, the steam-engine or water-wheel is spending its energy of mechanical motion to maintain it. Mechanical motions of large or small bodies are the antecedents of heat in every case. We have already considered how one kind of motion is capable of being changed into some other kind, also that such changes are not limited to large masses of matter, but are equally possible to the smallest. Impact upon a tuning-fork makes the latter vibrate on account of its elasticity. There are the best of reasons for believing that atoms and molecules are elastic bodies, and when struck in any way must vibrate like other elastic bodies, and must also have some rate of vibration peculiar to each element. This may be made clear to mechanically minded persons by considering what must happen to an elastic ring (Fig. 21) when its sides are pulled out and then let free. It will swing backwards and forwards first to a vertical ellipse, then to a horizontal ellipse, and so on. A ring six or eight inches in diameter, made of brass or steel wire, will show this kind of motion, which is vibratory as the result of mechanical impact. This is the kind of motion among atoms and molecules which one is to keep in mind as that set up among them when energy of any kind is spent upon them so as to produce heat. It is like the trembling motion of a bell when it is struck. Heat is this vibratory motion of atoms and molecules when hot. They do not necessarily have any translatory or oscillatory motion.

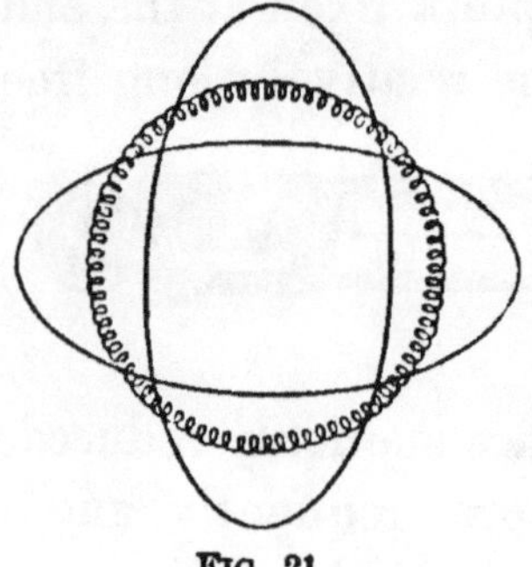

FIG. 21.

TEMPERATURE AND ITS MEASURE.

Our sensations of heat are not acute enough to enable us to determine by the feeling how much hotter one body is than another with any degree of precision. How little one's feelings can be relied upon may be learned by such experiments as the following. Fill two basins with water of the same temperature, and place one hand in each; to one hand it may seem warmer than to the other. Again, having three basins, — one holding water as hot as can be borne, the second, water ice-cold, the third, an equal mixture of the hot and the cold, — place one hand in the hot water, the other in the ice-cold, and let them stay a few seconds. Then dip both hands into the third basin; the hand that was in the hot water will feel very cold, while the hand that has been in the cold water will feel hot. In some fevers one may feel very chilly when he is really several degrees warmer than the natural temperature of the body. If, then, we are to measure differences in the warmth of bodies, we must depend upon something else than sensation. In reality we employ its mechanical effects, — its expansive power on either solids, or liquids, or gases. An instrument measuring this power is called a *thermometer* (a measure for temperature). The common form with which all are familiar consists of a glass bulb on the end of a long tube. Mercury fills the bulb and reaches a short distance into the stem at the lowest degree of cold to which it will be subjected. If this bulb be thrust into ice-cold water, the end of the mercury column will reach a certain point on the tube, which may be marked upon it; or, as is more commonly done, a scale plate may be firmly attached to the stem and the height of the mercury column be scratched upon it. If the whole be thrust into boiling water, the mercury will expand and fill the tube to a higher level; and if this place be marked on the scale, there will be the two fixed points indicating the freezing and boiling points of water. These are found to be uniform, and the mercury will move to these points respectively when placed in freezing or in boiling water. The space on the scale between these two fixed points is graduated in two different ways. In one, **F** (Fig. 22), the space is divided into 180 equal parts called degrees, and the lowest one is marked 32. The same spacing is carried still further down the scale, to 0 or below, where the numbers begin 1, 2, 3, and so on, and are read below zero. This makes the difference between 0 and the boiling point of water to be 212 such divisions or degrees. Such a scale is called Fahrenheit Scale. In the other way, **C** (Fig. 22), the space between the freezing and boiling points of water is divided into 100 equal parts called degrees, the freezing point of water being the zero of this scale. This is a very convenient scale for scientific work, and is called the Centigrade; but the other is in most common use, and will be employed in this book. With either of these thermometers it is possible to find the degree of heat in the air, in liquids, or in other bodies. Their scales are convertible from one to the other, for 180 divisions of the Fahrenheit are equal to 100 of the Centigrade, $\frac{180}{100} = \frac{9}{5}$. That is, one degree Centigrade is $\frac{9}{5}$ larger than the Fahrenheit; but their zeros do not coincide; therefore to reduce Fahrenheit degrees to Centigrade degrees, subtract 32 and multiply by $\frac{5}{9}$. To reduce Centigrade degrees to Fahrenheit, multiply by $\frac{9}{5}$ and add 32. For example: A Fahrenheit ther-

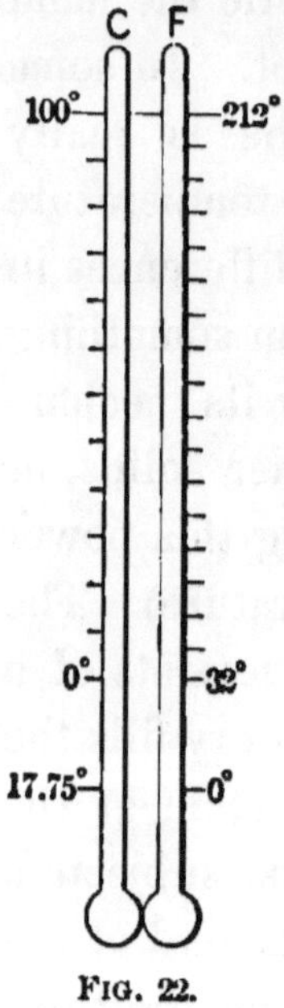

C F
100° 212°
0° 32°
17.75° 0°

FIG. 22.

mometer indicates 80°; what would be the indication on a Centigrade thermometer?

$$80 - 32 = 48,\ 48 \times \tfrac{5}{9} = 26.6°\ C.$$

What is the reading on a Fahrenheit scale when the Centigrade reads 25°?

$$25 \times \tfrac{9}{5} =, 45\ \ 45 + 32 = 77°\ F.$$

Sometimes thermometers are filled with alcohol instead of mercury, as it will not freeze at the temperature that will solidify the metal, but the scales are constructed in the same way. The following table gives some of the temperature ranges to be met with:

ABOVE 0° F.

Electric arc	6000°
Platinum melts	3400°
Bright red heat	1200°
Red heat just visible	1000°
Heat observed in India	140°
Human body	98.6°
Water freezes	32°
Mixture of ice and sal ammoniac	0°

BELOW 0° F.

Mercury freezes	—39°
Arctic cold	—70°
Artificial cold	—400°
Absolute cold	—459°
Temperature of space	—459°

Excepting the last two, these are observed temperatures. The last figure is called *absolute zero* to distinguish it from the other zeros. It is believed to be correct, for reasons that will be given further along in the book.

THERMODYNAMICS.

The temperature of a cupful of water would evidently be the same as that of the remaining water in the pail from which it was taken, but the amount of heat would depend upon the quantity of water; in two cups full there would be twice as much as in one. The temperature of a spark might be a thousand degrees, but it would not have heat enough to make an appreciable difference in the temperature of a pail of water if quenched in it. It is plain, then, that a distinction must be kept in mind between the temperature a body may have and the amount of heat it may have. Seeing that it will take twice as much heat to heat two pounds of water one degree as it will to heat one pound one degree, it has been found convenient to adopt a heat unit, which is *the amount of heat necessary to raise the temperature of a pound of water one degree.* When ten pounds of water are heated ten degrees, the water is said to possess 100 heat units, and if heated 100°, 1000 heat units. When a pound of water at boiling point, 212°, is cooled in any way to freezing point, 32°, it has lost 180 heat units.

It has been pointed out that friction results in heat, and if water be subject to friction it becomes heated. Churning it by revolving a paddle in it heats it appreciably. By arranging a paddle **p** (Fig. 23) in a known weight of water in **B**, and driving the paddle by means of a known weight W falling a given distance, one may know how much work is being done by multiplying the weight by the distance it falls; and by noting the temperature of the water at the beginning and at the end of the operation by the thermometer t, the rise in temperature equivalent to the work is known. Careful experiments of this kind have shown that the work

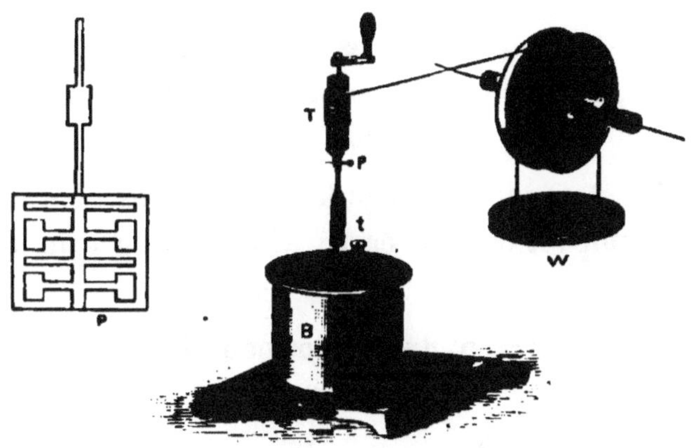

FIG. 23.

done by 778 pounds falling one foot, equal to 778 foot-pounds, will raise the temperature of one pound of water one degree. As the amount of heat necessary to raise a pound of water one degree is the heat unit, it follows that *778 foot-pounds is the mechanical equivalent of one heat unit* — a number to be remembered.

This means that the energy in a pound of water represented by one degree is equal to the energy represented by 778 foot-pounds, and if that water energy be properly applied it is capable of doing that amount of work.

Mechanical energy and heat energy are therefore

convertible quantities, that is, either may be changed into the other. This relation may be thus stated:

$$\text{Work} = 778 \times \text{heat units};$$

which is called the first law of thermodynamics. For instance, a hundred pounds of boiling water cools to 60°; how much work is represented by the loss in temperature?

$212 - 60 = 152° =$ loss in temperature,
$152 \times 100 = 15{,}200$ total loss of heat units,
$15{,}200 \times 778 = 11{,}825{,}600$ foot-pounds.

To what temperature would 100 pounds of water be raised by impact on the earth after falling a mile in a vacuum?

$5280 \times 100 = 528{,}000$ foot-pounds,
$778 \times 100 = 77{,}800 =$ amount of work needed to raise it 1°,
$\frac{528000}{77800} = 6.8° =$ rise in temperature due to the fall.

If the water were at 32° to begin with, its temperature after the striking would be $32 + 6.8 = 38.8°$, if all the work was spent on it.

FUELS.

We use wood, coal, coal oil, and gas for the sake of the heat that can be got from them when allowed to burn. When used for such a purpose they are called fuels. It has been found experimentally that the heating power of a given kind of fuel depends upon its amount, that is, its weight. When a pound of coal is burned in the air, it yields a product of carbonic acid gas, and generates heat enough to raise 14,500 pounds of water 1°, — that is, its heating power is 14,500 heat units. In like manner the heating value of a pound of the following substances has been determined:

Wood	7,000	heat units.
Wax	19,000	"
Phosphorus	10,350	"
Coal gas	22,500	"
Hydrogen	62,000	"
Fats	17,460	"
Coal oil	18,000	"
Bituminous coal	14,700	"
Anthracite	10,800	"
Pure carbon	14,500	"

These numbers are to be understood as representing the number of pounds of water that may be heated one degree by the combustion of one pound of the substance. For instance, a pound of wood when burned will heat 7000 pounds of water one degree; a pound of hydrogen will heat 62,000 pounds of water one degree; and so on for all the others. It should be remembered, too, that if a pound of wood will heat 7000 pounds of water one degree, it will heat 1000 pounds 7 degrees, or 100 pounds 70 degrees, for, *the product of the weight of water into its rise in temperature gives the heat units.*

One may now compute how much working power there is in a pound, or any other quantity, of any substance used for fuel when its heat unit value is known. Thus, the mechanical equivalent or working power of one heat unit being 778 foot-pounds, the foot-pounds of work a pound of wood when burned can do is equal to the product of the mechanical equivalent, 778, multiplied by 7000, the heat unit value of wood.

$$778 \times 7000 = 5{,}446{,}000 \text{ foot-pounds.}$$

For bituminous coal it is

$$778 \times 14{,}700 = 11{,}436{,}600 \text{ foot-pounds.}$$

Ten or a hundred pounds of either will of course give ten or a hundred times the amount of work one pound will give.

These constant numerical relations between heat and work have been verified in many ways, and now when one looks on a lump of coal he may be able to see something more than a mere lump of inert matter. Let him compute to what height the energy in the lump of coal would raise it if it were applied to such a purpose. In the table, fat is shown as having more energy to the pound than coal. Fat is one of the necessary ingredients of food. Indeed, nearly all foods may be used as fuel, and each particular kind has its own heating value. Bread and butter, if fed into a furnace, will make steam even better than coal; it is too costly for such purpose generally, but in regions where wood and coal have been scarce, corn has been used as a substitute. One may consider fuels as matter laden with a definite amount of chemical energy per pound, and such as are used for foods supply the body with the energy needful for movements and for its various functions.

Bearing this in mind, if one can know the heating value of the food he eats in a day, he may compute approximately the amount of work he can do in the way of climbing, shovelling, and so forth. When one becomes tired he has nearly used up his supply of energy, and as the bodily organs have a definite rate at which they can transform such food energy into work, they can be made to work but at a definite rate. This is very different with different individuals, and different races of men have very different natural rates of such bodily changes, which cannot be much increased by exercise.

The sources of heat are friction, condensation, percussion, chemical action, and all other forms of energy, and whenever heat appears, some other kind of energy has been spent to produce it. The amount spent is equal to that of the heat energy produced. We have already seen that energy $\frac{wv^2}{2g}$ and work pd are quantitatively related, and here we have a similar relation between heat and work. It is apparent that energy and heat are directly related when a definite amount of one is convertible into a definite amount of the other. Indeed, we shall presently see that it is only a transformation of the kind of motion we call mechanical or chemical into that vibratory kind among atoms and molecules that we call heat.

PHENOMENA OF HEAT.

I. **Expansion.** — 1. *Solids.* It is a familiar enough fact that heat expands bodies, and the action in thermometers illustrates it, whether the thermometers be made in one way or another. It remains to point out how it is that heat-expands. Consider a piece of wire an inch long, made of any substance. Its molecules are in the neighborhood of the fifty-millionth of an inch in diameter, and so fifty millions in a row make an inch. As the wire is a solid it possesses cohesion — the molecules stick tight together. Suppose these molecules be absolutely quiet, having no motion at all, and that they be crowded together so each one touches its neighbor like a row of coins or rings (Fig. 24). In such an arrangement the distance **ab** (Fig. 24) will be the shortest length into which this row may be crowded.

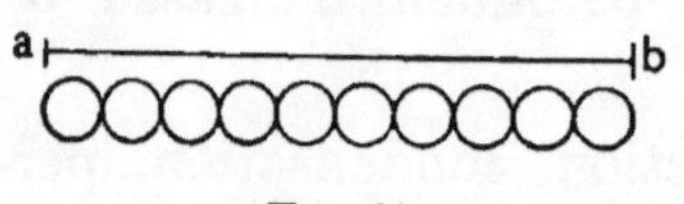

FIG. 24.

Suppose, however, that each molecule, by vibratory motion, changes its form, elongating and contracting on its axis; each one will then need more room for itself than it needed when at rest, and this will result in elongating the row (Fig. 25), and the measure **ab** will be too short, — how much too short will depend upon how great may be the amplitude of the individual vibrations multiplied by the number of molecules in the length measured. If the amplitude of such motion could be as great as half the diameter of each molecule, the increase in the length would be half the original length, and the inch of wire would become an inch and a half long. No such great change in length takes place in bodies, so no such relative amount of motion can take place in a single molecule; but if one can measure the increase in length of a piece of wire when heated, he can determine how much each molecule moves by

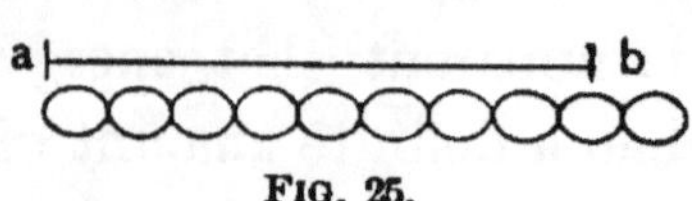

FIG. 25.

dividing the increase in length by the number of molecules in the length. In any case it is a very minute quantity, measurable in terms of the billionths of an inch, and in thousandths of the diameter of the atom.

The increase in length for many substances, due to a rise in temperature of one degree, has been most carefully measured, and is found to differ very much among different substances, but to be uniform for a given substance. The rate of increase for a unit length for one degree is called the coefficient of linear expansion of a substance.

If we call this increase in length for one foot for one degree a, then l feet will increase la for one degree, and for t degrees it will be lat; that is, the increase in length will be proportional to the length of the bar and to the temperature to which it is raised. Thus, the coefficient of expansion of wrought iron is .00000675, which means that a wrought iron rod one foot long, when heated one degree, becomes 1.00000675 feet in length; if one mile long it would become 1.00000675 miles, the increase being .0356 of a foot; if heated 40° the increase would be 40 times that, or 1.42 feet.

The following table gives the coefficient of linear expansion for some of the common substances:

Glass	.0000047
Cast iron	.0000062
Steel hardened	.0000068
Steel soft	.0000060
Copper	.0000095
Silver	.0000105
Lead	.0000155
Zinc	.0000163

Although all these numbers are very small, it may be noticed that they differ a good deal, some of them being two or three times greater than others, zinc having the greatest expansion rate.

The changes in length due to difference in temperature, while small for small bodies and for ordinary variations in temperature, become noticeable when either one is considerable. It would not be safe to lay railroad rails in coldest winter weather with their ends touching, for on some days in summer the temperature might be nearly a hundred degrees higher and the rails would be correspondingly longer, a difference of over three feet per mile, and in expanding they would crowd each other out of place and endanger travel. Tires for carriage wheels are heated four or five hundred degrees to expand them, then when placed in cooling they shrink and become very tight. A difference of the hundredth of an inch in the length of a common pendulum causes the clock to err about ten seconds in twenty-four hours, and a rise in temperature of 25° may produce this difference. To counteract this difference a glass cylinder filled with mercury is substituted for the ordinary disk. As the rod expands downwards the mercury expands upwards, and keeps the center of gravity at the same distance from the point of support. If two long strips of metals like iron and zinc be soldered together through their length, the strip will curve when heated or cooled, and thermometers are made on this principle.

CUBIC EXPANSION.

Bodies expand in breadth and thickness as well as in length; and in general, equally in all directions. The coefficient of volume, or cubic expansion as it is called, is ordinarily three times what it is for the linear expansion; thus for copper the cubical expansion is $.0000095 \times 3 = .0000285$. Some crystals expand unequally in their different dimensions; wood also expands and contracts more in breadth than in length. So long as there is no chemical change produced by heating or cooling, all bodies regain their original dimensions when brought to their original temperatures. This shows that the changes take place according to laws, and one may rely upon uniformity of action in this as in all other physical processes in nature.

2. *Liquids.* Like solids liquids change their volume when their temperatures change, only, on account of the slighter degree of cohesion among their molecules, their change in volume is considerably greater. Fill a glass flask (Fig. 26), holding a pint or more, with water, and, having a glass tube with a fine bore thrust through the cork; stop it tight and so that the liquid stands at some height in the tube; tie a thread around the tube to mark the place of the top of the water. The

Fig. 26.

warmth of the hands on the flask will be sufficient in a few seconds to make an appreciable rise in the tube. The same is true of mercury. There is a curious exception to the increase of volume on heating observable water. If water at the freezing point, 32°, be taken in for the above experiment, the volume will decrease and the top of the column in the tube will fall until the temperature rises to 39°, when it will begin to increase again. In this way it has been found that water is at its greatest density at 39.2°. Water at this temperature expands, whether it be heated or cooled. The cubical expansion of mercury and of alcohol are utilized in the making of thermometers, the cubical expansion of mercury being .00010 per degree, and of alcohol .00055.

3. *Gases.* As gases consist of molecules that do not cohere, but freely move about and bump against each other, it may be expected that the effect of heating the molecules will be to make them move more quickly, bump harder, and require more room for each molecule; so, if there be room, a volume of heated gas will beat back the neighboring gas and occupy more space, — that is, it will expand, and for a given degree of rise in temperature will expand very much more than any solid or liquid, for it has no molecular cohesions to counteract. If a given volume of air or other gas be taken at 32° and raised one degree in temperature, it is found that its volume has increased $\frac{1}{491}$ part. If it be heated two degrees its volume has increased $\frac{2}{491}$; and so on for any number of degrees, being always $\frac{t}{491}$, so if it be raised 491° its volume will be doubled.

In like manner, if the same volume at 32° be cooled 1° its volume will be diminished $\frac{1}{491}$ part, 2°, $\frac{2}{491}$, and so on down. Another way of stating this is to say that 491 cubic inches of a gas at 32° become 492 when raised 1°, 493 when raised 2°, and in every case the volume is $491 + t$, t being the number of degrees above 32°. In cooling, also, 491 cubic inches become 490 at 31°, 489 at 30°, and so on. These are results obtained by experiment, and this number 491 indicates that at 491° below the freezing point the gas would cease to exist, — not that the matter would be annihilated, but that it would be no longer in the gaseous state. This point of temperature, 491° below the freezing point of water, or 459° below the zero of the Fahrenheit scale, is called *Absolute Zero*, a term which means the total absence of any degree of heat. One may reckon the temperature of bodies in general on this basis, and then the temperature is called the absolute temperature. For instance, the absolute temperature of freezing water is 491°, of boiling water $459 + 212 = 671°$, of the human body $459 + 98.6 = 557.6°$.

The behavior of gases in thus expanding and contracting with reference to such an absolute zero has been formulated into a statement which may be remembered easily. *The volume of a gas is proportional to its absolute temperature.* This is called the Law of Charles. Its utility may be seen by an example. A cubic foot of air at 32° is heated to 100°, what volume will it now occupy?

The absolute temperatures of the two volumes are

as follows: that for 32° F. = 491°, that for 100° F. = 491 + 68 = 558°.

As the volumes will be proportional to these numbers we have 491 : 558 :: 1 : x = 1.14 cubic feet, nearly.

What will be the volume of air to which a cubic foot will be reduced when its temperature is lowered from 75° to 0°? The corresponding absolute temperatures will be 491 + 43 = 534 and 491 — 32 = 459. Then 534 : 459 :: 1 : x = .859 of a cubic foot.

How much air will escape from a room 20 feet square and 10 feet high when its temperature is raised from 40° to 70°? Volume of the room is 20 × 20 × 10 = 4000 cubic feet. Absolute temperatures of the two extremes are

491 + 8 = 499, and 491 + 38 = 529.
499 : 529 :: 4000 : x = 4240 = expanded air.
4240 — 4000 = 240 = cubic feet escaped.

II. Conduction. — If the end of a rod of iron be thrust into the fire, the end held in the hand will presently become too hot to be held, unless the rod be a long one; the heat slowly creeps along the rod, molecule by molecule. The process is called *Conduction*, for the molecules must be in cohesive contact in order that heat may travel from one to the other. But different substances have very different rates of conduction, silver and copper being very good, while German silver, wood, and paper are relatively poor. All the objects upon the table may be at the same temperature, but to the touch they may seem to be different; the better the conducting power, the colder the object will appear to be, as it robs the hand of its heat, conducting it away at a swifter rate than poorer conductors can. Furs and feathers, being poor heat conductors, serve to prevent the loss of bodily heat of animals and birds, while our clothing, varying with the season, serves the same purpose for us.

III. Specific Heat.—The specific heat of a substance is the amount of heat it takes to raise the temperature of a pound of it one degree, compared with the amount required for the same weight of water taken as unity. Thus a body requiring but half as much would have its specific heat = .5. The heat required for heating a pound of water one degree is called a heat unit, which has a mechanical equivalent of 778 foot-pounds. The specific heat of a substance is then the ratio of its heat unit per pound to that of water per pound.

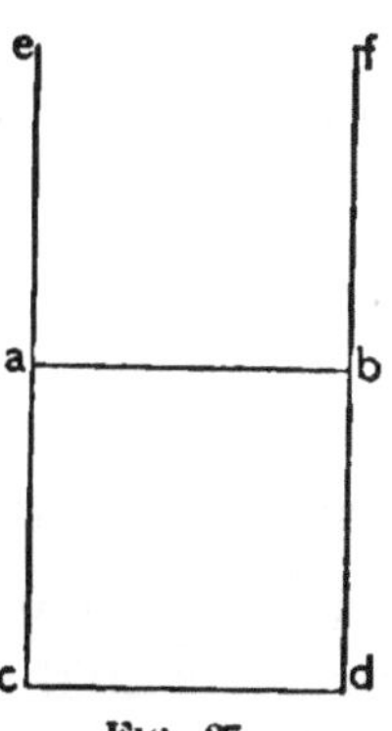

Fig. 27.

1. *Of Gases.* When a gas is heated, and thus its molecules have swifter free-path motion imparted to them, they strike upon the walls of the containing vessel with greater velocity and produce a correspondingly greater pressure; they have more energy. Suppose **ecdf** (Fig. 27) be a vessel in which the partition **ab** fits air-tight, but is capable of moving up and down freely. Let **acdb** represent a cube one foot on a side so as to contain a cubic foot of air at 32°, which is the temperature outside. The pressure will be the same on both sides of **ab**. If heat be applied to the

enclosed air while the partition **ab** is held firmly in place so that the air cannot expand, it will require a certain definite amount of heat to raise the temperature of the contained air one degree. But if the partition be permitted to move, the increased pressure due to heat will raise it somewhat, but in raising the partition it will be doing work against the pressure of the air; pressure upon the square inch being 14.7 pounds, upon the square foot surface it is $144 \times 14.7 = 2116$ pounds, and the amount of work *pd* will equal the product of 2116 into the distance the partition will be raised. It is found, however, that when by expanding the air does work, it requires more heat to raise its temperature one degree than when it does not expand, in the ratio of 1.41 : 1.

The ratio of the number of foot-pounds of energy required to raise a pound of the gas one degree in temperature, to 778:

(1) In an open vessel, doing work against air-pressure, is called *specific heat under constant pressure.*

(2) In a tight vessel, doing no work, is called *specific heat under constant volume.*

For specific heat under constant pressure it is .2375. For specific heat under constant volume it is .1674, and $\frac{2375}{1674} = 1.41$. For example: To raise the temperature of a pound of air one degree, when the air could not expand, would require .1674 as much as a pound of water would require in being heated one degree, or $778 \times .1674 = 130.2$ foot-pounds; when it could expand it would require 1.41 times as much: $.1674 \times 1.41 = .2360$, and $778 \times .2360 = 183.6$ foot-pounds.

This means that when a gas is heated in a tight enclosure where it cannot expand, all the energy is employed in heating the molecules; where expansion can take place, a part of the energy of the heated molecules is expended at once in doing work, and it therefore requires more heat to bring the temperature to the same point, and in the above ratio.

Whenever a gas does work it always loses temperature, as the energy for doing the work comes from the heat in the gas. The translatory motion represented in the work is the transformed vibratory motion of the molecules. It is this that makes possible steam-, gas-, and air-engines.

2. *Of Solids.* How much work in foot-pounds is needful to raise the temperature of a pound of water one degree has been shown to be 778, but no other substance requires so much. How much any given substance requires can be found by taking a pound of it, raising its temperature one hundred degrees, and then plunging it into a pound of water and observing to what temperature the water rises. Suppose a pound of mercury at 212° be cooled in water at 32°; the mixture is found to have the temperature of 37.9°. The water has, then, been warmed 5.9°, while the mercury has lost 174.1, and the ratio of 174.1 to 5.9 is as 1 : .033; that is, the amount of heat needed to raise the temperature of a pound of mercury one degree is only one-thirtieth that required for the same weight of water.

The specific heat of water is taken as the standard and is called unity, or 1. As it is higher than that of any other liquid or solid substance the others will be fractional.

The specific heat of a number of substances is given:

Water	1.0000
Ice	.4900
Iron	.1138
Copper	.0939
Silver	.0570
Lead	.0314

Specific Energy of Gaseous Molecules. — The heat energy of a molecule of hydrogen is the same as that of a molecule of oxygen when they have the same temperature, but hydrogen being only one-sixteenth as heavy must have sixteen times as many molecules to the pound. Hence a given weight of hydrogen will have sixteen times as much energy as the same weight of oxygen. In order, then, to produce a rise of temperature of one degree in a pound of hydrogen, sixteen times as much heat is needed as for the same weight of oxygen. The specific heat of hydrogen, then, is sixteen times that of oxygen. In general, the lighter the molecules of a gas the more numerous must they be in a pound of the substance, and the higher must be its specific heat.

Specific Heat of Solid Elements. — The specific heat of an element varies inversely as its atomic weight, or the product of the specific heat into the atomic weight equals a constant quantity which is about 6.4; thus the atomic weight of lead is 207, its specific heat is .0314 and $207 \times .0314 = 6.43$. If this number 6.4 be divided by the specific heat of an element it will give the atomic weight, and that is one way chemists employ to determine atomic weights.

The specific heat of a substance determines how high its temperature will rise when a definite amount of heat or of work is spent upon it. For instance, it requires 778 foot-pounds of work to be done upon a pound of water to raise its temperature one degree. If 778 foot-pounds of work were done upon a pound of iron, its temperature would be raised above that of water in proportion as its specific heat is lower; that is, $\frac{1}{.1108} = 9.02°$.

A lead bullet weighing an ounce strikes a target with a velocity of a 1000 feet a second. How high is it heated by the impact, if one-half of the energy is spent on it, the other half on the target?

$$\frac{wv^2}{2g} = \frac{\frac{1}{16} \times 1000^2}{64} = \frac{\frac{1000000}{16}}{64} = 976 \text{ foot-pounds.}$$

976 foot-pounds spent upon a pound of water would raise its temperature $\frac{976}{778} = 1.25°$; if spent on an ounce it would raise it $1.25 \times 16 = 20°$. If spent on lead having specific heat of .0314, its temperature would be raised $\frac{20}{.0314} = 636°$; but if one-half be spent on the target it will be $\frac{636}{2} = 318°$ = the rise in temperature of the bullet.

One may now by careful thinking understand how molecules of different kinds may have the same amount of energy in the form of heat when their temperatures are the same, for if their atomic weights are different they must have different degrees of amplitude of vibration, — that is, their velocities mûst be different; yet when equal weights of different elements are taken, they differ very much in their capacity for heat.

IV. Changes of State.—*Solid to Liquid.* The states of matter have been described as *solid*, *liquid*, and *gaseous*, and it is one of the properties of heat to change solids to liquids or gases.

That ice is frozen water and that it may be again changed into water is familiar enough to every one. Now that more definite knowledge is possessed of heat and its mode of action it becomes possible to explain how the changes are effected.

The distinction between a solid and a liquid is, that the molecules in the solid cohere so strongly together that they are not easily pulled apart, so that if one part of a solid body is pushed or pulled the pressure is felt by the whole body which will move as a whole if part of it is; this is not the case with a fluid.

Let one now consider the molecules of any solid, as, for instance, ice at the temperature of 31°. In a cubic inch of it there are a certain number of molecules, and each one has a definite position, rate, and *amplitude* of motion. If that amplitude be increased they will bump upon each other harder and will bound away from each other with greater velocity, and when they bound away so far on the average that they get out of each others *range of cohesion*, the molecules can be individually moved without pushing or pulling their neighbors—the body is no longer a solid. The heat which is applied to melt ice does this. Indeed, the increase of amplitude of vibration destroys the cohesion, so that when this reaches a certain degree any further increase of the energy results in changing the ice to water instead of raising the temperature. In similar manner, when water is reduced to 32° its molecules are on the boundary such that any further decrease of individual motion brings them within cohesion distance, and then the water becomes solid. What is true for water and ice is true for iron, lead, gold, rocks, indeed most bodies, the difference between them being chiefly in the temperature at which such change of condition takes place. When a solid becomes liquid the process is called *fusion*, melting, or liquefaction. When a liquid becomes a solid the process is called *freezing*, or congealation. For a given substance fusion and solidification take place at the same temperature; thus, ice fuses and water solidifies at 32°.

The melting point of some of the elements is given on page 5. If iron melts at 3000°, then iron cannot remain solid when the temperature is higher than that, and in like manner when the temperature is below 32° water becomes ice. If, as is probable, the earth was once hotter than 3000°, there could not have been anything solid upon it. The waters of the oceans would have existed as steam. The lava that flows from volcanoes is melted rock and implies a temperature of two or three thousand degrees. On the other hand, in the polar regions the ice never melts; and on the dark side of the moon, where the sun never shines, the temperature is probably a hundred degrees below zero, or more.

As a general thing the colder a body is the stronger is its cohesion; it takes a greater pull to break a wire at 32° than at 50°, and still greater to break it at 0°. This increase in cohesion is found experimentally to hold good down to —300°.

Influence of Pressure on Fusion. — It has been found that pressure upon a body changes its fusing temperature, usually raising it. If the heat applied to melt a body has to do work against pressure, one would expect that more heat would be required ; in the same way and for the same reason, more is required to raise the temperature of air when it can expand and do work than when it does not have to do so. Thus, if it requires a temperature of 2000° to fuse common rock material in the air, to melt it when subjected to a pressure of some tons per square inch may require 2500°.

The deeper one goes into the earth the warmer it becomes, increasing at the rate of about one degree in each 60 feet. At that rate the temperature at the depth of 25 miles would be above 2000°, quite sufficient to melt rock if it were on the surface where the pressure is at most but about 15 pounds per square inch. But as one goes down into the earth the rock-pressure increases also, and much more rapidly than the temperature, the result being that the rocks cannot be melted, no matter what the temperature. This means that the earth is solid to the center, and not, as has sometimes been thought, liquid in the interior, with a relatively thin crust on which we live.

Liquid to Gaseous. The difference between the liquid and the gaseous state is the difference between very slight cohesion and none at all. If, therefore, the temperature of a liquid be raised so as to increase the average distance apart of all the molecules — altogether beyond the range of cohesion — they will fly off and become free roving gaseous molecules, and will maintain this condition so long as such temperature be kept up. The vibratory heat energy imparts translatory mechanical energy to the molecules and they bound away. This makes the difference between the liquid and the gaseous state — simply the amount of energy present among the molecules. After ice has become water, the molecular cohesion is very small, yet it has some value; but as the temperature is raised the cohesion becomes less and less, until at 212° it reaches the limit, where the molecules no longer cohere, but become free and gaseous. In doing this they necessarily occupy much greater space, so that a cubic inch of water becomes nearly a cubic foot of gaseous steam. Sometimes it is called a vapor, but there is no distinction between a gas and a vapor. The free paths of such molecules are now something like 250 times their own diameters. Such rapid increase in volume implies vibratory amplitude of molecular motion so great as to quite destroy cohesion among the colliding molecules.

Dissociation. This process can be carried on another step by increase of temperature. Water molecules are made up of two atoms of hydrogen combined with one atom of oxygen cohering together. If the amplitude of vibration of the individual atoms in the molecule be increased by increasing their temperature, it will presently reach a limit at which the atoms are no longer able to remain together, — they separate; that is, the water molecule is broken up, and the atoms become individual gaseous particles. This happens at about 4500°. This process is called *dissociation*.

Evaporation. Consider the surface of water in a saucer. It is in contact with the air above it. The air particles are bumping upon the surface molecules, but no water molecule is thus struck downward continuously with the same degree of impact. The surface molecules of the water are continuously being bumped from below upwards, for they are in practical contact with other molecules except above, and any such bump from below upon a surface molecule, occurring at an instant when there is no corresponding downward bump of an air particle, may cause that water molecule to be quite knocked away from the surface, and so become a free gaseous molecule. The less frequent the number of bumps from the air downward, that is, the rarer the air, the greater the number of water molecules which will be driven from the surface. This process of changing a liquid to the gaseous form is called *evaporation*. It is going on all the time from the surfaces of liquids and many solids, and at all temperatures, for even ice will evaporate. The wet clothes upon the line may freeze at first, but will presently dry, if left alone. A lump of camphor, if left in free air, will slowly be evaporated, and the metal mercury likewise, though at a very low rate; in any case, evaporation depends upon the temperature.

Solidification. At very high temperatures, such as that of the sun and of an electric arc, every kind of substance we are acquainted with is reduced to its gaseous form; so, on abstracting the heat from such bodies, they assume the solid form again, for the free path of their molecules is shortened until the molecules are within cohesion distance. Most of the chemical elements are to be found in the solid state; but hydrogen, oxygen, and nitrogen, as well as common air, require that the temperature be very low, and that pressure be applied in order to reduce them to the liquid or solid form. In this particular they do not differ in quality from other substances except in degree, — they assume the gaseous condition at a lower temperature. At the temperature of —360° both oxygen and common air have been solidified, and as they are magnetic, they will stick in masses to the poles of a magnet, as iron will do, and slowly evaporate as they become warmer.

It was stated on page 79 that when hydrogen and oxygen unite to form water they give up an immense amount of energy in the form of heat. Obviously, they cannot give up energy which they do not possess; hence one must think of these gases as possessing a great, but definite, amount of energy at ordinary temperatures, energy which is not heat, but may be transformed into heat.

It is thought by some to be probable that, if these and all other gases could have their temperatures reduced to absolute zero, they would cease to be gaseous in form, and their molecules would sink to the ground like so much dust. This conclusion follows from the consideration that it is heat alone that keeps any molecules in the gaseous state.

Crystallization. A large number of substances assume some regular form on becoming solid from either the liquid or gaseous state. Such symmetrical forms are called *crystals.* Ice is such a crystalline form, as may be seen in the fern-like shapes upon the window panes on a cold day, and snowflakes have a hexagonal form with six points or six sides in many varieties (Fig. 28).

FIG. 28.

More than a hundred have been pictured. Sulphur, alum, and sugar are substances that easily crystallize. Quartz, diamonds, and most minerals are frequently found crystallized.

A strip of zinc thrust into a beaker or bottle containing dilute lead acetate will be covered in a few minutes with a dense growth of lead leaves looking like vegetable forms. A drop of a solution of ammonium chloride spread upon a strip of glass and looked at through a magnifying glass will be seen to assume an arrangement like the streets of a city, and a solution of barium chloride will have the appearance of a lot of small bushes. If projected upon a screen by means of the *porte-lumière* and a beam of sunlight (p. 230), they may be seen to great advantage by a room full of people.

The ability to assume such regular shapes is inherent in the molecules, and depends largely upon the temperature, for at a definite temperature liquids will dissolve only a certain quantity of a substance, and if the temperature be made lower some of the substance will become solid in the crystallized form. Solid iron will crystallize if it be jarred often, which shows that the molecules are all the time under some sort of constraint, which tends to set them in symmetrical order, and jarring helps them to assume their more stable positions.

Cooling. There are two methods by which a body may lose its heat, — by *conduction* and by *radiation.* The first implies *contact* with a cooler mass of matter, and the second a *transformation* of the heat energy into ether wave energy, sometimes called radiant energy. In this place we are concerned only with the former. A heated body, if left to itself, will cool to the temperature of surrounding bodies; it will not get any colder. Out of doors the temperature in summer may be 70° or 80°. In winter it may fall to zero or lower.

The conditions equalize the temperatures of all bodies exposed to them, except such as are provided with some other means for changing them. If one blows his breath upon the back of the hand it will feel cool, although the breath is several degrees warmer than the hand. If a drop of water be spread over the hand it will feel cool, and if this be blown upon, it will be still cooler. All this happens because evaporation is going on. In order that the moisture upon the hand shall become gaseous, it must have heat from some source; the hand supplies it, and the feeling of coolness results. Blowing upon it increases the rate of evaporation. A drop of alcohol or of ether placed upon the bulb of a thermometer will lower the temperature of the bulb several degrees.

In like manner, when solids become liquids heat is abstracted from the surrounding body. Ice and salt, both solids, will melt if mixed together, and in becoming liquid will abstract heat from any available source. Water may be frozen by enclosing it in such a mixture. This is the common arrangement for making ice cream. Salt water does not freeze at 32°; at how much lower temperature it may remain liquid depends upon the amount of salt dissolved. With such a mixture the temperature may fall to 0° F.

Many substances will dissolve in water more rapidly if the water be heated. If the water be not heated the solution will go on at a slower rate, and the temperature of the mixture will fall. Thus, add some ammonium chloride to an equal volume of water at ordinary temperature, and stir it with the bulb of a thermometer. The temperature will fall 15 or 20 degrees. The heat needed for dissolving it comes from outside, from the hand or air or both, so the reduction of the solid to the liquid form is a cooling process for neighboring bodies.

When air is condensed it is heated, as shown by the experiment on page 71. If it be allowed to expand again at once, it regains its original temperature; but if it be allowed to stand after condensation until it has cooled to the temperature of surrounding things, and then permitted to expand, its temperature falls, and water in contact may be thus frozen. More heat is needed to maintain air as a rare gas than as a denser one.

Surface Tension. The fact that the surface molecules of liquids are subject to cohesive attraction only at their sides and underneath enables them to cohere together stronger than those beneath the surface do; the energy which is divided in every direction by one beneath the surface is divided among a smaller number on the surface, and is therefore stronger. This gives to the liquid surface a tension, acting much like a thin skin, always tending to contract the surface to the smallest dimensions. A soap-bubble shows this, for if blown and left on the bowl of the pipe it will contract so much as to enter the bowl. The globular form of a drop of water is due to the same condition. The surface tension of a liquid serves to restrain evaporation, so that it cannot go on at as swift a rate as it would if the temperature alone controlled it. Pure water has so high a surface tension that bubbles cannot be blown with it; but a little soap in it reduces the tension very much, and increases the rate of evaporation as higher temperature would do. Almost anything water will dissolve acts to thus reduce its surface tension.

Fill a saucer with pure water and scatter some lycopodium powder or dust on its surface. A drop of ether on a glass rod held an inch or two above will cause a lively scattering of the particles, as if repelled by the rod. The ether vapor is denser than the air, and falls to the surface of the water; it is at once absorbed by the water lessening its surface tension, and the part of the surface not thus affected pulls the particles away.

A few small crumbs of camphor dropped upon the surface of pure water will move about in a surprising way as if alive. The camphor is dissolved by the water at the points of contact, and this reduces the tension at those points, and the changes in tension result in pulling the bits about at a lively rate.

A drop of any of the essential oils, such as those of cinnamon, clove, or creosote, when dropped on a water surface, spreads about and assumes characteristic forms which enable one to identify these oils, as each has its own characteristic tension and cohesion.

The Boiling Point. — Whenever heat energy is applied to a liquid in any way in sufficient quantity so that beneath its surface the molecular cohesion is rapidly destroyed, bubbles are formed by the gaseous molecules which rise to the surface and, if free, escape into the air. If the liquid be water, and it lie in an enclosed vessel like a boiler from which the gaseous molecules, called steam, cannot escape, the pressure will rise, and the steam with its temperature and pressure may be conveyed by pipes, and used to warm houses or drive engines.

The boiling point of water is 212°, the same temperature at which steam will again condense to water on removing some of its energy. Other liquids have other boiling points; thus, alcohol boils at 140°, ether at 63°, mercury at 630°, lead at 2700°.

Pressure affects the boiling point of liquids very much as it affects the fusing point of solids, — it raises it. If a flask of water at 200° be placed under the receiver of an air-pump, and the air-pressure be removed, the water will boil at as lively a rate as if still over a fire. At high elevations, where the air-pressure is less, water boils at a lower temperature than 212°. In the city of Mexico, 7470 feet above the sea-level, water boils at 199°. In Quito, South America, 9340 feet high, it boils at 195°, and on the top of Mt. Blanc, 15,630 feet, at 182°.

If we have steam and water at 212°, it is evident that there is much more energy in the steam molecules than there is in the water molecules, for the steam has energy of free-path motion in addition to its internal vibratory rate that constitutes its temperature; this extra energy that the molecules of steam have, when their temperature is the same as that of the water from which the steam was formed, is called its *latent heat*. It means only that, in becoming water again, it will give out the same amount of energy in the form of heat, or that the mechanical energy of the steam molecule, represented by its free-path motion, will be changed into heat energy when it is again allowed to assume the liquid form.. In like manner, when water at 32° becomes ice at 32°, it gives up a relatively large amount of energy, the same in amount that was needed to change it from ice to water. This, too, has been called latent heat, but it signifies only a change in the form of the energy, and not that there is heat in the molecule which is latent. If a pound of ice and a pound of water, each at 32°, be exposed, in two similar vessels, to the same source of heat, at the moment the ice is melted in the one the temperature of the other will be 174°; showing that it takes as much heat to melt a pound of ice as it does to raise a pound of water from 32° to 174°, that is, 142 heat units. In like manner, if a pound of water at 212° be mixed with a pound of pulverized ice or snow at 32°, when the latter is melted the mixture will have the temperature of only 50°. The ice will have gained 18°, while the water will have lost 162°. Here again 142 heat units have been consumed in changing the ice to water.

When steam at 212° is converted to water at 212° it yields up energy which takes the form of heat.

A pound of steam at that temperature becomes water at the same temperature on giving up 966 heat units.

See, then, the energy measured in heat units which is imparted to a pound of ice in raising it to steam.

From ice at 32° to water at 32° =	142	heat units.
From water at 32° to water at 212° =	180	"
From water at 212° to steam at 212° =	966	"
Total,	1288	"

The number 1288 × 778 = 1,002,064 foot-pounds of

energy spent on one pound of ice, and this quantity it will give out as heat on cooling down to ice again.

VII. The Working Power of Steam.—On page 106 it is pointed out how much energy in foot-pounds is spent in raising a pound of ice to steam at 212°. At this temperature the pressure of the steam is the same as that of the air, namely, about 15 pounds per square inch. If water is boiled while exposed to the air the pressure never rises higher than this, but if the water is enclosed in an air-tight vessel or boiler, and heat still be applied, the pressure rises, and reacting upon the water stops the boiling until more heat be supplied; so when heat is abundant the temperature of the boiling water rises so that at 260° the pressure is 20 pounds above atmospheric pressure, at 308° it is 60, and at 365° it is 150 pounds per square inch above the 15 pounds pressure of the atmosphere.

Let it now be remembered that when a pound of water is raised one degree in temperature it has been endowed with 778 foot-pounds of energy. As the specific heat of steam is one-half that of water, if a pound of steam be raised from 212° to 300° — 88° it has had $\frac{778 \times 88}{2} = 34,232$ foot-pounds of working power added to it. The working power of steam is in its *pressure*, and so long as pressure is maintained by heat, so long can work be done with it.

The Steam-Engine.—The steam-engine is a machine for utilizing steam-pressure. The pressure of the steam upon the piston-head in the cylinder (Fig. 29) depends upon the temperature of the steam, and the area of the piston-head upon which the steam presses. Thus, if the area be one square foot, and the pressure 50 pounds per square inch, the total pressure will be $50 \times 144 = 7200$ pounds, and if the piston moves forward a foot with that pressure it does 7200 foot-pounds of work. In order to find the *work* the engine is doing, multiply the area of the piston in square inches by the pressure per square inch shown by the steam-gauge, and that product by the number of feet the piston-head travels per second or per minute. Thus, if the length of stroke be two feet, and if the balance wheel goes around twice a second, the piston has to move backward and forward twice in that time, that is, 8 feet; and if the area of piston be one square foot and pressure be 50 pounds per square inch, as above, then the work done will be equal to $7200 \times 8 = 57,600$ foot-pounds per second, which equals $\frac{57600}{550} = 104$ horse-power.

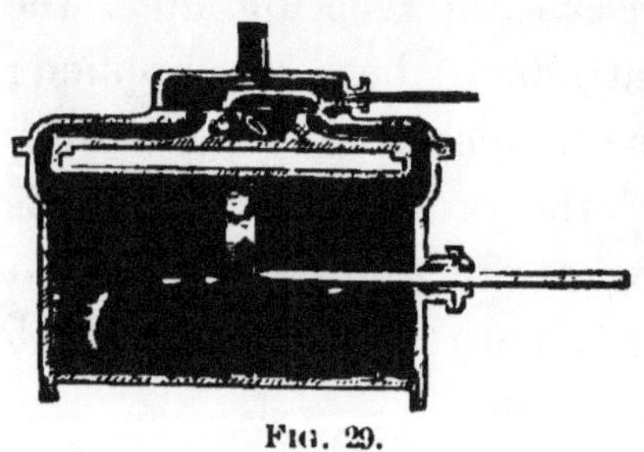

FIG. 29.

Because the steam has done work it has lost temperature proportional to the work done, so when it escapes from the engine it is no longer as hot as it was when it came from the boiler. If one knows the temperature of the steam as it enters and escapes after doing its work he can calculate how much energy it has lost; but this temperature must be measured on the *absolute scale.*

Suppose an engine employs steam at the temperature of 260°, and that it escapes into the air at a temperature of 215°; it has lost 260° — 215° = 45°. But 260° Fahrenheit = 260° + 459° = 719° absolute temperature, and 215°F. = 215 + 459 = 674° absolute.

The temperature at the start being 719°, and at the end 674°, there has been used 719 — 674 = 45°, which is only a little more than 6% of the whole amount.

If t represents the absolute temperature of the steam in the boiler, and t^1 the absolute temperature on its escape from the cylinder of a steam-engine after it has done its work, then $\frac{t-t^1}{t}$ represents the *efficiency*. This may be applied to any kind of heat engine.

The ratio of the amount of energy used to the whole amount supplied is called the efficiency.

If the steam-engine could be worked down to the temperature of 32°, its efficiency would be increased very much, for, as before, $\frac{719-491}{719}=\frac{228}{719}=32\%$, and if it could be worked down to absolute zero, the efficiency would be 100%.

As steam condenses to water at 212° its available gaseous pressure is gone, hence its serviceability increases as its temperature increases above that. It has not been found practicable to use steam at a higher temperature than about 400°, 859 absolute, as the oils needed for lubricating are decomposed by higher temperatures, and the valves leak. At that temperature the pressure is not far from 250 pounds per square inch.

If an engine be worked between the limits of 400° and 212° (859 — 671 = 188 absolute temperature) its efficiency would be the highest, namely $\frac{188}{859} = 22\%$. This shows that our present steam-engines are not economical, for under the best conditions three-fourths of the energy supplied to them is wasted.

The energy for making steam is supplied by the fuel, — coal, wood, gas, or oil. Suppose it be coal, one pound of which when burned can raise the temperature of 12,000 pounds of water one degree. One pound, then, has $12{,}000 \times 778 = 9{,}336{,}000$ foot-pounds of work possible in it. A horse-power is 550 foot-pounds per second, or $550 \times 3600 = 1{,}980{,}000$ per hour; hence one pound of coal has energy enough to maintain a horse-power $\frac{9{,}336{,}000}{1{,}980{,}000} = 4.7$ hours, if it could be all utilized. As a matter of fact it takes about two pounds of coal to maintain a horse-power for an hour in good engines, and two or three times as much in the poorer ones. Two pounds of coal have $2 \times 9{,}336{,}000 = 18{,}672{,}000$ foot-pounds' work. In the engine they give but 1,980,000, hence $\frac{1{,}980{,}000}{18{,}672{,}000} = 10\%$, which represents the real efficiency of a *good* steam-engine.

The *steam plant*, as it is called, consists of a *furnace* for combustion, a *boiler* for converting heat energy into mechanical pressure, the pipes for directing the pressure to the engine where the steam-pressure is transferred to the moving *piston*, and then to the *wheels*, from which it is transferred by belts or gears to other machinery.

We start with chemical energy in cold fuel, which is *transformed* into heat in the furnace; the heat is transferred to the water, and changes it into a gas with free-path motions and pressure measurable in pounds per square inch. The pressure on the moving piston gives *work*, and its *rate* is measured in horse-power. From the beginning to the end of the process the machinery, of whatever kind it is, does nothing but change the energy from one kind or direction to another; and no more energy can be obtained from the engine than goes in at the furnace. This, perhaps one would say, was obvious enough, yet there are not a few persons who imagine that mechanism can of itself produce energy. This is what is implied in all attempts at what is called *perpetual motion*. The effort is to make some machine do work which is not supplied with energy from some external source. No one has succeeded hitherto, and we know why not. A machine transforms or transfers energy supplied to it, nothing more.

Nature of Heat. — From all that has been said one may conclude that heat phenomena are due to the *vibratory motions* of atoms and molecules — not translatory in either long or short paths. Translatory motions are the *results* of heat motions, for a molecule might have any assignable temperature, and yet remain in the same place, if not otherwise disturbed (p. 72). Also that the energy of heat is only energy in this atomic and molecular form of vibration. When this *form* or kind of motion is changed into another in any way, it is no longer heat. When one considers the characteristic of heat, it is proper to consider it as a *mode* of *motion.* When one considers what and how much work a definite amount can do, it is proper to consider it as a *mode* of *energy.* When one would think out what happens in a mechanical way in heat phenomena, one must think of modes of motion, — of the kinds of motion that precede the appearance of heat as well as those into which the heat loses its identity. The word heat itself gives no clue to its nature or mode of operation.

QUESTIONS.

1. How many heat units are there in ten pounds of each of the following substances: coal, wood, kerosene oil, fat?

2. If ten per cent. of the coal were ash or stone, how many heat units would there be in 100 pounds?

3. How many foot-pounds of work are the equivalent of one ounce of pure coal?

4. If the work-power of one pound of pure coal were applied to raising itself, how high would it be raised?

5. What weight of coal would be necessary to raise yourself ten feet high, if all its working power could be utilized?

6. If only ten per cent of its power could be thus used, how much more would be needed to do the same work?

7. If a steam-engine does 2,000,000 foot-pounds of work in an hour, what is its horse-power?

8. If a steam-engine utilizes but 6 per cent of the working power of the coal and burns a ton an hour, what horse-power has the engine?

9. A large ocean steamer uses five hundred tons of coal a day. If it takes two pounds per hour per horse-power, what is the horse-power of the engine?

10. If it takes two pounds of coal to maintain a horse-power for an hour in a given steam-engine, what is the *efficiency* of the

engine, that is, what per cent of the whole energy of the coal is made serviceable?

11. How many degrees above the freezing point will 100 pounds of water be heated by all the heat developed by burning one pound of pure coal?

12. How many degrees will the same amount be heated by burning one pound of wood? Also by a pound of coal oil?

13. If it takes thirty times as much heat to raise the temperature of a pound of water one degree as it does to raise a pound of lead one degree, how many degrees will a pound of lead be heated by the work of 778 foot-pounds?

14. If the specific heat of iron be .11, how many degrees will a pound of iron be heated by the amount of work that will heat a pound of water one degree?

15. A bullet weighing one ounce strikes a target with the velocity of 1000 feet a second;

(1) How many foot-pounds of energy does it have? $\frac{wv^2}{2g}$=energy.

(2) If the bullet were made of water, how many degrees would it be warmed by the impact?

(3) If the bullet were made of lead, specific heat = $\frac{1}{30}$, and half the work were spent upon the target, how much would the bullet be heated?

(4) If the bullet were iron, how many degrees would it be heated under the same conditions?

CHAPTER IX.

ELECTRICITY AND MAGNETISM.

ELECTRICITY.

I. Origin of Electricity. — If a strip of zinc be put into some water in a tumbler or other convenient vessel, nothing appears to happen to it. If a little acid of any kind be added to the water, bubbles will collect upon the zinc, indicating that the zinc is being dissolved slowly. Now let a carbon rod such as is used for the arc light be put into the same liquid, but not so as to touch the zinc — the carbon is not acted upon apparently, and the zinc is dissolved slowly as before. Let the carbon lean so as to touch the zinc, and at once the zinc shows signs of increased activity, — the bubbles collect fast on the carbon, break away from it, and rise to the surface. This action stops when the zinc and carbon are separated. If a wire be connected to the carbon rod and the other end of it be touched to the zinc, the action is again set up; the zinc is more rapidly dissolved when it is in connection with the carbon either directly or by means of the wire. This indicates that in some way the carbon affects the zinc through the wire. It makes no difference what the kind of wire may be. Copper wire is more generally used because it is abundant, easily bent or twisted, and is otherwise well adapted to electrical work; but iron, German silver, platinum, and other metals show the same phenomena, indicating some sort of physical action going through the wire. The action in the solution is partly chemical and partly some other kind, which has the name *electrical* to distinguish it from the first; and the action that takes place in the wire is called an *electrical current.*

That something goes on in the wire is shown in other ways also. For instance, if the wire conveying such electrical current be held close over a common compass needle and parallel with it (Fig. 30), the needle will be deflected one way or the other, depending upon which end of the wire is connected with the zinc, for if the

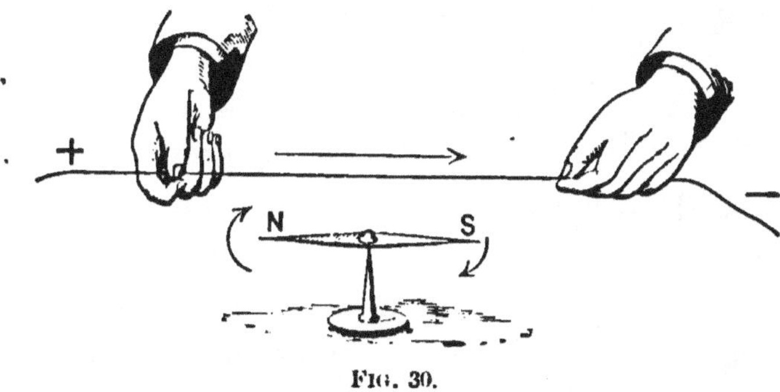

Fig. 30.

ends of the wires be changed from one to the other element in the solution the current will be reversed in the wire and the needle will be deflected in the opposite direction. For experimental purposes such an electric cell as the above will not answer as well as some of those found in the market, but they differ only in the quantity of current given, and we may use any of the numerous forms of Leclanche battery cells in place of the arrangement described above. The Leclanche (Fig. 31) consists of zinc and carbon in a solution of sal ammoniac, and, as before, the zinc is dissolved, the carbon is not, and a current of electricity traverses the wire when it connects the two substances.

It has been found needful for convenience to assume that the current in this wire has always a certain direction which depends upon the carbon element, *going out from* that, and one may with a compass needle quickly determine in which direction the current is going in the wire. If one bring the wire down *over* the needle and *parallel* with it, as in Fig. 30, if the current be *going north, the north end will turn to the west.* If going south, the south end of the needle will turn to the west. Remember this for its convenience.

FIG. 31.
Leclanche Cell.

A compass needle with a coil of wire around it or under it through which a current of electricity may be sent is called a *galvanometer* (Fig. 32).

If it be a delicate one the needle will move when the ends of a copper and an iron wire connected with it are dipped into a drop of salt or acid water. The combination forms a small battery.

FIG. 32.

Another way of generating electricity is to take an ordinary magnet, either a straight bar or of horse-shoe form, and wind about one of its ends a dozen turns of copper wire so loosely that the coil will slip off and on easily. Then if the ends from this coil be connected to the galvanometer (Fig. 33) and the coil be moved back and forth over either end of the magnet the needle will move this way and that, depending upon which way the coil moves and also upon which pole is used. But the galvanometer must be so far from the magnet that the movements of the latter will not appreciably disturb it.

FIG. 33.

The effects upon the needle are the same as was noticed with the chemical or galvanic cell, indicating that a similar condition exists in the connecting wire, that is, that a current of electricity is going through the wire. If, instead of moving the coil back and forth, it be held still, and a piece of iron or the keeper of the magnet, as the armature of it is called, be pulled off its poles and put on again, the effects are the same; the direction of the needle is reversed when the action is reversed, showing the existence of opposite currents.

In one important point this way of generating an electric current differs from the chemical. In the latter, the current continues as long as the wires are connected to the battery. In the former, the current continues only as long as there is some kind of movement of the coils at the ends of the magnet; when the movement stops the current stops, although the wires may be in connection. The first is called a *continuous* current, the second an *intermittent* current.

A third way of generating an electric current is to have two wires of different metals, as iron and copper, twisted together at one end (Fig. 34), and heated at this *junction.* A continuous current through the galvanometer is furnished by this means, but with wires as described it is not as strong as the current from the cell. If such elements as antimony and bismuth be taken in short bars and soldered together at their alternate ends, there is formed what is called a *thermo pile*, which may be used for some purposes. A still better one is made by using bars half an inch thick and two inches long, of an alloy of antimony and zinc for one metal and German silver for the other, all arranged in a circular form, as shown in Fig. 35, so that a gas jet can heat the ends of the elements; twenty such elements in one gas jet are better than the chemical cell described. There are still other sources of electrical currents, such as electrical eels and fishes, atmospheric conditions which produce lightning, and physiological conditions, for every time a muscle is contracted a slight current of

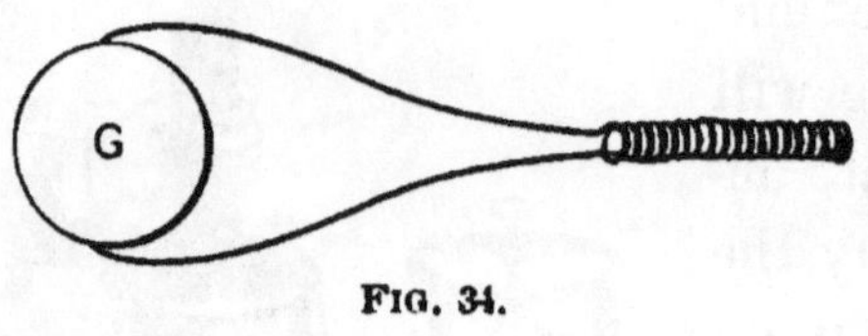

FIG. 34.

FIG. 35.

electricity is generated, so slight that the most delicate means are needful to observe it.

Let us now return to these various sources described, and study them in the light of what has preceded this chapter.

When the magnetic needle turns this way or that in response to the electric current in the wire, it is to be remembered that such movement indicates *energy*, for it would not move unless it received somehow and from somewhere a pressure — and pressure always implies energy. When the direction of the current is changed the pressure upon the needle changes too, and as this evidence of pressure ceases when the wire is disconnected from the source of action, one must look to see what evidence there is of activity in these various sources.

1. *In the Galvanic Cell.* The solution dissolves the zinc; a solid is made to assume a liquid form, and, as was pointed out on p. 94, wherever there is a change of form of matter from solid to liquid or gaseous, there is always an exchange of energy. Zinc has more atomic energy in its solid form than it has when chemically combined with anything, the same as hydrogen and oxygen have, for when they unite they give out a large amount of energy in the form of heat. When combined with sulphuric acid and forming zinc sulphate a pound of it yields 3000 heat units of energy, but when this process takes place in a galvanic cell *this energy shows itself in a current of electricity* instead of heat. Hence, in a galvanic battery the source of the electric current is the molecular energy

of the zinc. The process we call a chemical process, and the product is zinc sulphate and an *electric current*, just as the product of the combination of oxygen and hydrogen is water and *heat*.

The energy of the electric current is distributed through the wire, and it possesses a property of pushing or pulling upon bodies in its neighborhood such as heat energy does not have. The magnetic needle is peculiarly sensitive to such action of a current, and it is, therefore, employed as an indicator or measurer of it.

2. *In the Magnet and Coil.* When the magnet and coil of wire are at rest — that is, when no mechanical energy is employed to disturb their positions — there is no electrical current and the needle remains still. Unless some work is being done with them, there is no energy to be spent by them in moving the needle. The antecedent to the current in this case is mechanical energy, as chemical energy was the antecedent in the galvanic cell. The product is the same but the factors are different.

3. *In the Thermo Pile.* The thermo pile may be connected to the wire and galvanometer, but the needle is as idle as if it were not connected until one face of the pile is heated; then a current at once results and the needle moves. We are supposed to have a definite idea of heat as a form of energy in which the atoms and molecules are vibrating at high rates. As long as they are all at the same temperature they are in equilibrium, but when one junction or one face of the pile is heated higher than the other,

energy is being spent at different rates upon the two metals that form the couple. Suppose the two elements of the thermo pair be copper and iron; they have different atomic weights and, therefore, have different vibratory rates and different rates of heat conduction. Where the elements are in contact, and are heated, their rates are quickened and they mutually interfere with each other. If the other ends be connected with a wire so as to form a circuit, a part of this energy shows itself as a current of electricity, and the rest is spent in raising the temperature; if the wires are not so connected, the whole energy is spent in raising the temperature. As in the other two sources so in this, energy has to be spent in order to produce the current, and the energy of the current thus derived shows itself in moving the needle. But in this case the kind of energy supplied is unlike either of the others. In the first it is chemical; in the second, mechanical; in the third, heat. The product is the same, — an electric current in each circuit of wire. We have presented a curious matter for consideration, — like results from seemingly very unlike causes.

In every case energy of some kind has to be spent in order to produce electricity. In the atmosphere, when lightning appears — which is a transient current of electricity — there are always energy transformations going on. The wind and clouds and rain, and oftentime hail, show that mechanical causes, heat causes, and chemical causes are all present.

II. Electrical Terminology. — It is as needful to have some standards for use in electrical phenomena as it is in mechanical or heat phenomena. In mechanical measurements we use pounds and feet and quarts; in heat we use heat units and degrees of temperature. If electricity possesses energy it has pressure, for, as explained already, whenever energy is acting there is pressure, and the unit of electrical pressure is called a *volt*, just as the unit of mechanical pressure is called a pound, and whatever produces electrical pressure is called electro motive force.

The unit of electrical current is called an *ampère.* It means the rate at which a current is flowing in a wire; very much as the rate of the flow of water from a pipe is indicated by cubic feet per second. A current of one *ampère* will in one second give a *quantity* of electricity called the *coulomb.* That is, a coulomb is an ampère per second.

The size of a conductor as well as its quality determines how much electricity can go through it in a given time, very much as the size of a pipe, its diameter, length, and smoothness of bore will determine the amount of water that can go through it under given pressure. These conditions of size offer a resistance to the flow of water called friction and similar conditions of hindrance to the flow of electricity are called electrical resistance, and a unit of such resistance is called an *ohm.*

The unit of electrical power is called the *watt*, and represents the work equivalent of a current of one ampère when the electric pressure is one volt.

The unit of electric	pressure	is the	volt ;	its symbol is	*E*.
" " " "	current	" "	ampère ;	" " "	*C*.
" " " "	resistance	" "	ohm ;	" " "	*R*.
" " " "	quantity	" "	coulomb ;	" " "	*Q*.
" " " "	power	" "	watt ;	" " "	*W*.

These factors stand in certain algebraic relations to each other which may be thus expressed :

$$\frac{\textit{electric pressure}}{\textit{electric resistance}} = \textit{electric current},$$

or, using the symbols in place of what they signify,

$$\frac{E}{R} = C,$$ which is known as *Ohm's Law.*

This is used in the following way.

The pressure in an electric circuit is 10 volts, its resistance is 5 ohms ; what current is flowing in it?

$$\frac{10 \text{ volts}}{5 \text{ ohms}} = 2 \text{ ampères}.$$

Again, a current of 8 ampères flows in a circuit having

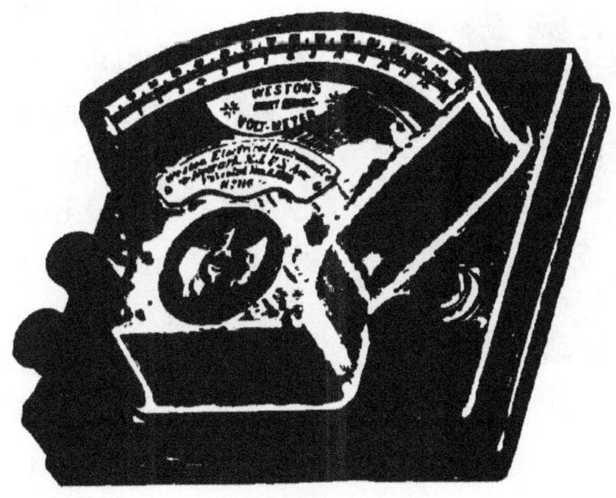

FIG. 36.

a known resistance of 15 ohms, what is the pressure? $RC = E$, $15 \times 8 = 120$ volts.

If any two of these factors be known the third one may be computed in this simple way.

In order to measure these quantities there are various

Fig. 37.

kinds of instruments. Formerly galvanometers were almost universally used, but now instruments called voltmeters (Fig. 36) and ammeters (Fig. 37) are in use which indicate directly the value of the pressure or the current, just as weighing balances indicate weight, and thermometers the temperature, without any calculations.

Fig. 38.

It will be assumed that such instruments are used in the indicated experiments which follow; also that there are boxes containing wires of known resistance called *resistance boxes* (Fig. 38) with conveniences for connecting wires, and for changing the resistance. Such wires are only measured lengths of German silver wire which have resistances so arranged that one can use any number of ohms from one to the upper limit of the particular box.

Electrical Measurements.

I. Electric Pressure. — To determine *voltage* or *pressure.* Suppose there be for study two Leclanche cells.

(1) Connect one of the cells to the voltmeter with wires two or three feet long that have been coiled up by winding them around a convenient rod to keep them from sprawling (Fig. 39). Be sure that the metallic connections are good, and also that each wire is connected to the proper terminal on the voltmeter. The pointer will indicate the voltage of the cell. For such a cell as the above the pressure will probably be somewhere in the neighborhood of 1.4 volts. See whether both cells have the same pressure.

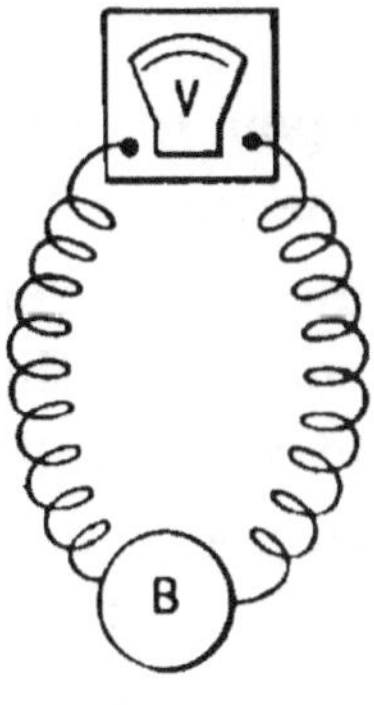

FIG. 39.

(2) Couple the two cells together; first, the zinc of one to the carbon of the other, and the remaining carbon and zinc to the voltmeter, and note whether or not the indicated voltage is the sum of the two. Second, couple them zinc to zinc and carbon to carbon, and these wires to

the voltmeter. If there be other kinds of cells, such as copper sulphate or bichromate of potash, they may be compared with the others. Ordinarily copper-sulphate cells have 1.1 volts each, and bichromate-of-potash cells about 2 volts. The voltmeter is so constructed that only a very weak current goes through it, and only a minute quantity of electrical energy is used.

Now let a short piece of wire be touched to the terminals of the battery while the voltmeter is connected, and observe the latter (Fig. 40). When the wire first touches, the needle falls a good deal and continues to fall while you look at it.

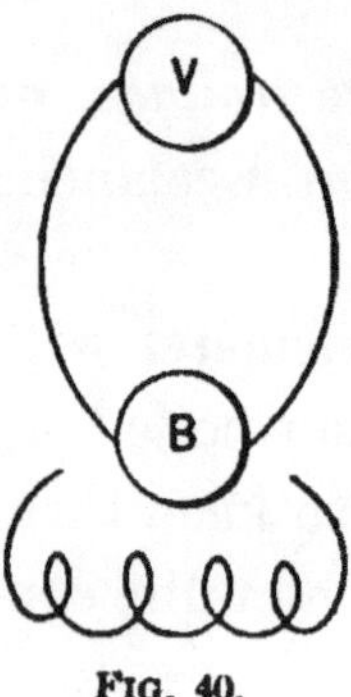

FIG. 40.

Actions of the Cells. — The explanation of this will be seen in making a cell like the first one mentioned (p. 114) of zinc and carbon in a solution of dilute sulphuric acid. The bubbles will collect upon the carbon rod and nearly cover it, thus keeping the liquid from touching it. The bubbles are hydrogen gas there set free by the decomposition of water. Of course, if the liquid cannot touch the carbon the latter may as well be out in the air; the current is nearly stopped, and the electric pressure is reduced as indicated. This collecting of the gas upon the carbon is called battery *polarization*. The zinc first decomposes the water and unites with the oxygen forming zinc oxide, which then combines with the sulphuric acid forming zinc sulphate. In chemical symbols it is like this:

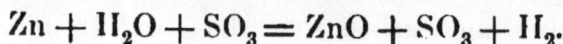

$$Zn + H_2O + SO_3 = ZnO + SO_3 + H_2.$$

The ZnO combining with SO_3 becomes $ZnSO_4$, and that leaves the hydrogen free, but here is the most singular phenomenon in the whole process, — none of the hydrogen appears at the zinc surface when the water is decomposed; it all is at the carbon plate. To understand this it would be needful to look down among the molecules with the imaginative eye to see what must be going on. As the zinc is capable of combining with oxygen in definite proportion, there is what is called chemical attraction between the two. When the zinc is immersed in the solution the attraction is not quite strong enough to decompose the water, as a piece of potassium will do, but the attraction is there, and every water molecule adjacent to the zinc is swung round so that its oxygen face fronts the zinc. It can do this, for, as has been pointed out, there is no friction among water molecules; they are free to turn about in any direction. So the oxygen sides are towards the zinc and, of course, the hydrogen sides are away from it. The carbon does not act upon the water, it will not decompose it except at a red heat; and when a current of electricity flows, as it always will in such a cell, it goes from the *zinc to the carbon*, and there is an interchange of atoms among the molecules along the whole line. Fig. 41 shows this, where the water molecules are represented by circles, one above the other, the hydrogen circles having lines drawn across them, the lower circles being oxygen. If the lower left-hand oxygen atom 1 be taken away,

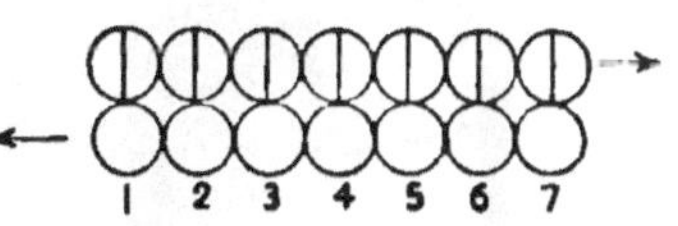

FIG. 41.

the hydrogen with which it was combined at once combines with **2**, the oxygen of the next adjacent molecule, and its hydrogen to the next, until the last at the surface of the carbon where there is no longer any oxygen for hydrogen of **7** to combine with, and therefore it is set free. There has been decomposition and recomposition of every molecule along the whole line.

This is the way the hydrogen gets to the carbon in the cell, and as it sticks there it presently brings the cell to a standstill. To prevent this various means are employed. In the Leclanche cell the carbon is mixed with binoxide of manganese, the whole is then pressed into sticks or cylinders and baked. The manganese gives up freely some of its oxygen to the hydrogen just set free, and water is again formed at the carbon surface. This gets rid of the gas, but the solution left is so weakened after a time that the cell is spoiled. Another way is by putting the carbon into a porous jar containing nitric acid or chromic acid, and setting this down into the solution with the zinc. This answers very well while the strong acid lasts. When the zinc is put into dilute sulphuric acid such a cell is called the *Bunsen cell* (Fig. 42). It has a pressure of about 1.9. If the zinc be put into sal ammoniac, instead of into sulphuric acid as proposed by the author, the pressure is 2.2 volts. Such batteries

FIG. 42.

are called two-fluid cells, and the second liquid employed, *viz.*, the one in which the carbon stands, is called a *depolarizer*, as its chief function is to prevent the collection of hydrogen gas on the plate immersed in it.

While the cell being examined is connected with the voltmeter, slowly raise the elements out of the liquid, observing whether there is any change in the voltage. It will be found that the extent of surface exposed to the liquid does not affect the pressure. That is, the pressure depends upon the *character* of the chemical work and not its amount. A small cell of a given kind has the same electric pressure as one made ten or a hundred times larger. This may remind one of the pressure of water per square inch, which depends upon the depth of the water rather than its quantity.

II. Measure of the Current.— Let now the ammeter **A** be put in the circuit with the battery **B** and the set of resistance coils **R** (Fig. 43). As the resistances are varied the current becomes more or less, according to Ohm's Law. Suppose that the ammeter indicates one-half an ampère ; the voltage being known, suppose it be 1.4 ;

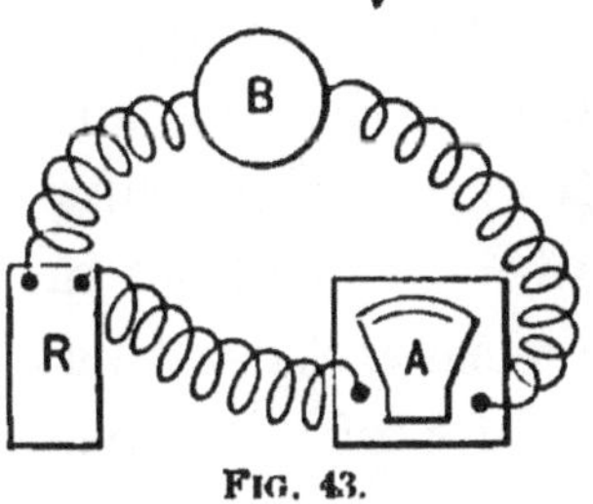

FIG. 43.

then according to Ohm's Law $\frac{V}{C} = R$, $\frac{1.4}{.5} = 2.8$ ohms.

This resistance is made up of the resistances of the battery itself, of the connecting wires, of the ammeter, and the resistance in the box of coils. Usually the resistance of the ammeter is so small it need not be reckoned, and the same with the connecting wires, so the **R** is made up of the resistance in the battery and in the coils. That in the coils can be read, suppose it be 2 ohms. Then as the total resistance is 2.8 ohms, the resistance of the battery itself will be .8 ohms. This resistance of the battery cell depends upon the *size* of the cell. Raise the elements out of the solution slowly while the motion of the ammeter needle is watched; the current becomes less and less as the immersed surface is lessened. If the surface in contact with the solution be thus reduced one-half, the resistance of the cell, if measured as above, will be found to be 1.6 ohms.

III. Measure of Resistance. — The resistance of a length of wire of any kind may be measured by putting it in circuit with the ammeter and a battery cell, and noting the current that is indicated. Now replace the wire with the resistance box and see what number of ohms must be put into the circuit in order to give the same reading upon the ammeter. This number of ohms will be the resistance of the wire being measured. This is called measuring resistance by *substitution*. There are other ways of measuring it with a higher degree of precision than this, but this way seems to show the nature of the process, and still better, it shows that electrical phenomena are as subject to law as mechanical or heat phenomena, and may be as definitely known.

IV. Energy of a Battery. — The working power of water is measured by multiplying the water pressure by the amount of water that flows per second; and in like manner the working power of electricity may be found by multiplying its pressure in volts by its current in ampères, that is, EC, and this product is called *watts*, of which 746 equal a horse-power, and each horse-power equals 550 foot-pounds per second (p. 40). If a battery has one volt pressure and gives one ampère current, its energy is $1 \times 1 = 1$ watt $= \frac{1}{746}$ of a horse-power, $\frac{1}{746}$ of $550 = .75$ of a foot-pound per second. This is a small quantity. How may it be increased?

Suppose the cell has one volt pressure and a resistance of half an ohm, and the external wire through which the current has to flow has half an ohm, the whole may be represented thus: $\frac{1}{.5 + .5} = 1$ ampère. If we make the cell larger we make its resistance less, but that does not change its pressure. Suppose it be made five times as large, the resistance of the cell will be reduced to $\frac{1}{5} \times \frac{1}{2} = \frac{1}{10}$ of an ohm, and the current will be $\frac{1}{.1 + .5}$ $= 1.66$ ampères, and the working power will be 1×1.66 $= 1.66$ watts, or 1.24 foot-pounds per second.

Suppose the pressure of the cell be increased by employing different chemicals. This may be done so as to make it equal two volts. With the other conditions as before this will double the working power, for now there will be $2 \times 1.66 = 3.32$ watts. It is not at present practicable to have much more than two volts to a cell, neither is it practicable to have the resistance of a cell less than .1 of an ohm; so the above shows about the maximum work to be got out of one cell of ordinary size.

If one wishes to find how many cells are needed for a horse-power he must remember that watts is the *product* of volts and ampères ; 746 watts are required. If the current be 10 ampères there must be 74.6 volts, and if each cell has two volts, $\frac{74.6}{2} = 38$ cells — as there cannot be half a cell. If a current of 20 ampères be used the number of cells will be one-half, but their size will necessarily be greater, for it will be remembered that the current of a cell depends upon its size. One can see from the above why galvanic batteries have not been used for many purposes even when only a small amount of power is required. Another reason that they are not available for such purpose is the necessity for frequent renewal of the materials, which, without the greatest painstaking, is sure "to make a mess."

Again, zinc has the same function in a battery that coal has in a steam-engine. Zinc has 3000 heat units value per pound, while pure coal has 14,500.

$778 \times 3000 = 2,334,000$ foot-pounds for the pound of zinc.
$778 \times 14,500 = 11,281,000$ " " " " " coal.

The coal has nearly five times as much as the zinc. Zinc is worth about 100 dollars a ton, while coal can be had for 3 dollars a ton. A given amount of power got from zinc costs over a hundred and fifty times as much as it does from coal. It should be kept in mind that the substances used in batteries have definite working power per pound and that no kind of machinery or combinations can increase it but may greatly waste it.

V. Electrical Conductivity. — If one take a number of different kinds of wire such as silver, copper, iron, German silver, or others, *all of the same length and thickness*, and connect them one at a time to a battery or other source of electricity, having an ammeter in the circuit, he will see that the current will not be the same for any two. For silver the current will be the greatest, and it will be much less for German silver. The differences are due to the degrees of *conductivity* of the substances of which the wires are made. If, for instance, the current shown by the ammeter when the silver wire was tried was 1 ampère, while for iron it was but .16 of an ampère, and for German silver only .07 of an ampère, it would show that silver was more than six times as good a conductor for electricity as iron, and fourteen times as good as German silver. In this way tables of conductivity have been made of all the metals and some other substances. Here are a few to show the order of differences in them, silver being the best.

Silver.	Platinum.
Copper.	Iron.
Gold.	Lead.
Aluminum.	German silver.
Zinc.	Mercury.

Liquids are still poorer in conductivity than any of the metals, pure water being very nearly a non-conductor; a drop or two of acid or a pinch of salt in a pint of water improves its conductivity thousands of times. Bodies whose conductivity is very small are called non-conductors; thus wood, glass, and gases are non-conductors. Conductivity and resistance are complementary terms, that is, one increases as the other diminishes. This is sometimes expressed thus: conductivity varies inversely as the resistance. For most electrical purposes resistance is the term used rather than conductivity.

These differences in ability to conduct electricity possessed by the different metals depend upon their molecular properties, very much as heat conduction does. Indeed, bodies that are good conductors of one are good conductors of the other.

So far the degree of conductivity has been considered between different metals when they are of the same length and diameter.

A copper wire the twenty-fifth or .04 of an inch, in diameter, and 150 feet long has a resistance of about one ohm. If two such wires were stretched together, the two could conduct twice as much electricity in a given time as one could do, that is, the conductivity of the two is twice that of one. If both these were made into one wire of the same length, the *cross section* of it would be twice as great as the cross section of one. If the conductivity be doubled the resistance is reduced to one-half; hence we say that resistance *varies inversely as the cross section of the conductor.* When the diameter of a wire is doubled, its cross section is increased four times, for the areas of circles are to each other as the squares of their diameters. Hence a wire 150 feet long and .08 of an inch in diameter will have a resistance of only *one-fourth* of an ohm.

Doubling the length of a wire doubles its resistance, so we say the resistance of a wire *varies as its length.*

Putting these facts together the law may be briefly stated thus:

The resistance of a wire varies as $\frac{l}{d^2}$, where l is the length and d its diameter; also it varies with the kind of material, k; altogether it varies as $\frac{kl}{d^2}$.

Tables of the sizes of copper and iron wires and their resistances per unit of length or weight are very common, and are constantly referred to in electrical industries. See table, page 312.

The resistance of a common telegraph line is 10 or more ohms to the mile, that of an arc-light wire about 2 ohms to the mile. Hence a telegraph line 100 miles long will have a resistance of a 1000 ohms or more; and an arc-light circuit 20 miles long 40 ohms. These figures are for the wire alone.

All electrical devices have more or less resistance. A telegraph *relay* may have 150 ohms' resistance, a *sounder* 10, an incandescent lamp 100, an ocean cable 3000.

The conduction of electricity requires substantial contact between the molecules of the conductor, in the same way as heat does. As heat separates molecules, it increases the resistance of a conductor. Conductivity increases as the temperature is made lower, and experiments indicate that at absolute zero of the heat scale none of the metals would have any resistance, they would all be perfect conductors.

MAGNETISM.

If two or three feet of insulated wire of any convenient size be wound about a piece of iron, as a large nail or bolt or plain rod (Fig. 44), and a current of electricity be sent through the coil, the iron becomes a *magnet*, and other pieces of iron will be attracted by it and stick to it if permitted. As soon as the current stops, the magnetism mostly disappears, and returns again on completing the circuit. If the same thing be done with a piece of hardened steel, the latter does not lose its magnetism on opening the circuit, — it has become a *permanent magnet*. It is more convenient to study the properties of magnetism with permanent magnets than with temporary or *electro-magnets*, as they are called.

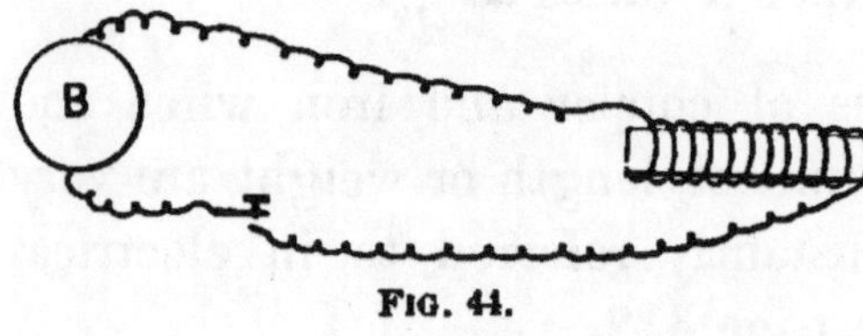

FIG. 44.

Magnets may be of any form; ordinarily they are straight bars or crooked like the letter U.

Suppose we have a straight steel bar magnet six or eight inches long, and an ordinary magnetic needle, which is also a permanent magnet, suspended so as to be free to turn towards any point of the compass. When the bar magnet is at a distance the needle points towards the north, and the end that points north is generally marked in some way to indicate it. If the bar magnet be brought into the neighborhood of the needle the latter will turn one of its ends toward the bar magnet. If an attempt be made to bring it nearer the other end of the needle that end will move vigorously away, and can be brought near only by forcibly preventing the needle from turning round. To describe these two actions, one is called *attraction*, the other *repulsion*. If the other end of the bar be taken in place of the one first tried, it will be found that the end of the needle which was attracted will now be repelled. If the bar be suspended by a string, so that it also may be free to turn, it too will point to the north; and if the end which is towards the north be marked in any way so that it may be recognized, and then the experiment be repeated, it will be found that *like poles of a magnet repel each other, while unlike poles attract*, but both poles will attract iron.

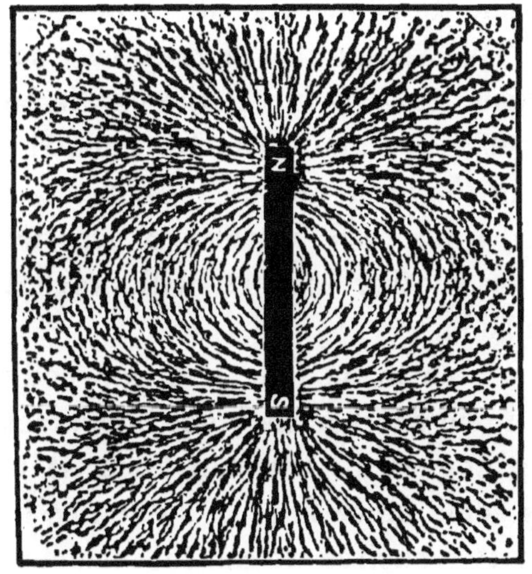

Fig. 45.

Now over the bar magnet put a piece of pasteboard or a pane of glass, and upon it scatter loosely some iron filings; jar the glass or paper a little, and the filings will be seen to arrange themselves in curved lines running from one pole to the other (Fig. 45). When thus arranged the pasteboard or glass may be gently raised and the arranged filings may be taken away. If the paper upon which the filings are strewn be first covered by a thin film of wax or paraffin, it may afterwards be warmed without disturbing the arrangement. The filings will sink through, and when cool again will be fixed. Such a thing is called a *magnetic phantom.*

Theory of Magnetism. The study of this phenomenon of magnetic arrangement has led to the theory of both magnetic and electric phenomena which we must here stop to consider.

In the first experiment mentioned above, the needle was moved this way or that by the *mere presence* of the bar magnet without one touching the other, or even being very near it. This shows that a *body may act upon another body without touching it.* If the needle be suspended in as perfect a vacuum as can be made, the action is not stopped in the least, and no substance has been found which, if interposed between a magnet and a piece of iron or another magnet, will prevent this attraction or repulsion.

As no one has been able to imagine how one body could in any way act upon another body not in contact with it, and with nothing at all between them, it is assumed that there is some other kind of substance in space than ordinary matter, such as we call the elements and their compounds. It is called *the ether.* It is believed, (1) that it quite fills space, (2) that it is not made up of atoms like ordinary matter, (3) that it is homogeneous — every part exactly like every other part, (4) that it is highly elastic, greatly exceeding steel in this particular, (5) that it is frictionless, so bodies can freely move through it and not lose their motion, (6) that the whole universe of matter, suns, planets, and stars are swimming in it — indeed, that it is illimitable, (7) that it possesses an immense amount of energy in various forms, and (8) that it is capable of transmitting energy of certain forms with enormous velocity. Light is one form of this energy, and is transmitted by this ether with the velocity of 186,000 miles a second. The ether must extend to the sun, for it takes eight minutes for the sun's light to get to the earth, a distance of 93,000000 miles. It must extend to the fixed stars, the nearest one being so far away that it takes 3½ years for its light to reach us; while some of the remote stars are so distant that their light requires thousands of years, traveling at the above rate, to get to us. As the light is a kind of energy and will do work, it must be energy while on the way, and energy is the product of something into a rate of movement. The rate we know; the something in this case we call *ether;* it cannot be matter.

It is through the agency of this space-filling ether that the magnet becomes able to act on another body. It acts first upon the ether about it, and the ether reacts upon the second body. Experiments have shown that this action upon the ether, which is called a *stress*, extends to an indefinite distance from the magnet in every direction, only becoming weaker as the distance increases. The bar magnet acts strongly upon the needle when near to it, at the distance of two or three feet much less; at ten feet away perhaps one could not observe that the needle has any motion, yet if the latter were very delicately poised it would move appreciably. There is no reason to doubt that the magnet affects the whole ether, only at great distances it is too weak to be observed. *The space about a magnet, in which it produces a stress, is called a magnetic field.*

The ether stress about the magnet is of such a nature that a piece of iron or steel or another magnet in the stress is simply *turned into a new position.*

Take a small compass and move it about the larger magnet, observing the position the small needle takes in different places. Then compare these positions with the lines of iron filings as shown on the phantom. They are the same in direction. The filings serve to show the direction of the stress in the field. Again, the lines all start from one of the ends or poles and go out in a curved line towards the other pole to reënter there, so that each line appears to represent a circuit, part in the magnet itself, the rest in the space about it. A *magnetic circuit* consists of the magnet itself and its field; the magnet may have any form. If it be shaped like the letter U, the field is close to the poles. Place a U magnet under the plate of glass, and sprinkle iron filings upon it. They will arrange themselves between the poles in straight and curved lines, and the closer the poles are together, the *denser* is the magnetic field.

When two similar magnets have their similar poles together, their fields are strengthened. When their opposite poles are together, they cancel each other. This can be seen by allowing one magnet to pick up a piece of iron, as a nail. Bring down slowly another similar pole upon the pole holding the nail; it will only stick the tighter to the magnet. Bring down the other pole to the same place and the nail will drop off easily, showing that the magnetism has been neutralized like two opposite pressures in equilibrium. Yet each magnet has lost nothing, for when separated each will attract and repel as well as ever. The reason for this apparent loss is simply that the *field* has been destroyed because all the stress that before was in the space is now in the opposite magnet. In other words, iron and steel are better conductors of magnetic stress than the air is, and the magnetism goes through them when they are present. A piece of iron in a magnetic field absorbs the stress and is turned by it into a new position, and every piece of iron that has absorbed the stress is itself a magnet.

Take a board nail and some tacks. Unmagnetized, the nail will not pick up a single tack. Slowly bring down towards one end of the nail one pole of a magnet. Before the magnet has touched the nail, the latter will be found to have become a magnet, and a tack will stick to it. If now it be permitted to touch the magnetic pole, other tacks will stick to the first tack, and if the magnet be a strong one, half a dozen tacks will hang together end on end, even when the magnet is separated slowly from the nail. This action by which a piece of iron becomes magnetic by being in a magnetic field is called *magnetic induction.*

Let a bar magnet of any convenient size be put into a shallow dish and covered with water to the depth of half an inch; scatter a *few* iron filings upon the water, not close together. They will not sink, but will arrange themselves with reference to the magnet. Now let another magnet be brought over, and a few inches from the filings, and turned so as to present first one pole then the other to them, the filings will be seen to turn about each time a different pole is presented to them, showing that the *particles are themselves magnets*, and vary their positions to accommodate themselves to the magnetic field they chance to be in.

Magnetize a knitting-needle or a piece of watch-spring by drawing its ends over the ends of a permanent magnet. If tested, it will be found to have poles, and to behave like other magnets. Break it in two in the middle and test each end; each piece has its poles. Break each piece again and again as long as is practicable; each fragment is a complete magnet, and there is every reason for supposing that if this process could be carried on till the iron molecule itself was reached, it would be found to be itself a magnet. This is believed to be the case. If it be true, then when such a molecule is in a position where it is free to turn about, it will do so if a magnetic field acts upon it with changing polarity.

Fill a test tube three-fourths full of iron filings. If it be held near the poles of a suspended magnetic needle, every part of it will be attracted by the needle; and as the needle is free to turn, either pole will point to any part of the tube. Now hold the tube against the two poles of a strong magnet and gently jar it so as to allow the filings to arrange themselves more easily. Remove it from the magnet without shaking it, and test for magnetism as before, and the tube will be found to have poles, and will behave like any other magnet, only it will be weaker. Shake up the filings and again test it, and all its polarity will be gone. The *arrangement* of the parts has been broken up. While it was a magnet all the like poles of the individual particles faced the same way. When it was shaken up they were all disarranged so their individual fields neutralized each other; there was no longer any field common to all. If a magnetic phantom, such as was described, be carefully made and fixed upon a piece of paper and suspended horizontally by a thread so as to be free to turn, it will face north and south like any other magnet, and remain a magnet for an indefinite time.

All this means that the difference between a piece of ordinary iron and a magnet is that the molecules of the latter all face one way, while in the iron they face every way. A current of electricity in a coil makes the molecules of iron all face one way, and so does another magnet. If they can be fixed in such positions, the piece remains a magnet; if they are not so fixed, the magnetism is apparently lost. A piece of soft iron has its molecular cohesion of position so strong that the molecules return to their original positions as soon as they are free to do so, that is, as soon as the induction pressure ceases. A piece of steel has carbon disseminated through it which, when it is hardened, prevents the molecules from so easily changing back to their original positions, so it remains a magnet.

From this one is to learn that we do not make iron molecules magnetic; they are already magnetic. We

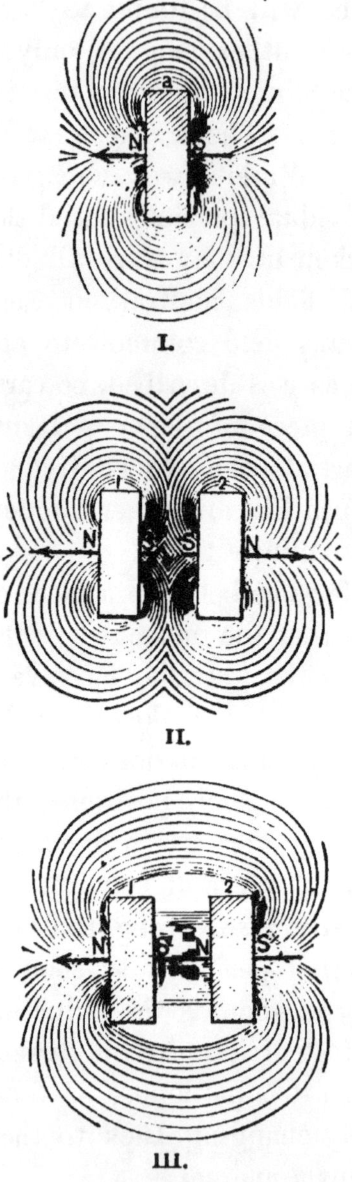

Fig. 46.

arrange them so their magnetic fields shall not neutralize each other. The accompanying diagrams will help one to understand magnetic arrangement.

In I let **a** represent a single iron molecule — a magnet, with a north and a south pole, **N** and **S**, and a field represented by the lines coming out on the **N** side and reëntering on the **S** side.

If II be two similar ones turned so that the **S** poles are adjacent, all the lines of each one will go from **S** to **N**, as indicated by the arrows, the same as in I, and these opposite movements between them producing opposite stresses, tend to make them separate from each other. This is called *repulsion*.

In III the difference is only that **2** is turned round so its **N** side faces the **S** side of **1**. The lines of **1** will now go around **2** and enter in its **S** side and go straight through both. The field of each is enlarged, and the ether stress now acts so as to crowd the two together. This action is called *attraction*. So if a thousand or a million are similarly arranged their fields are correspondingly increased, and the pressure of the fields acts in one direction or the other to produce attraction or repulsion.

Earth's Magnetism. — There are many interesting experiments which may be tried with magnets, but they are changes rung upon the preceding ones. There is one, however, well worth repeating that emphasizes some of these principles. The magnetic or compass needle points in the direction it does because the earth as a whole is a magnet, and has a field which acts inductively upon all other magnets in it. Take a rod of iron two or three feet long — a piece of gas pipe, for instance — and bring it near the poles of a free-turning needle. Hold the rod horizontal, and at right angles to the poles of the needle; either pole will be found to attract it, showing it has no polarity of its own. Now bring the rod over the north pole of the needle so as to be nearly vertical, but two or three inches from the needle. The bar will now repel the needle; and if the bar be struck with a hammer while in this position, the needle will move promptly round, showing the decided polarity of the bar, which will strongly attract the south end of the needle. Reverse the bar, and the polarity of it will be reversed; and if the bar again be moved to a horizontal position, it loses all polarity, and will be equally attracted by both poles of the needle. Thus the earth makes a magnet of a piece of iron held properly in its magnetic field, and the polarity of the lower end of it is north.

The magnetism of the earth is believed to be due to the large amount of iron in it; indeed, there is reason for thinking that the most of the interior of the earth is made up of that element. We live in a magnetic field, in which the lines representing it go from one pole to the other, high through the air and beyond the air.

Magnetize a disk of sheet steel as large as a half dollar so that its poles will be on its edge and opposite each other, then make a filings phantom as before, and the lines will show how they are distributed above the earth's surface. The pillars of light seen in auroral displays keep parallel with these magnetic lines of the earth's field.

Effects of Heat on a Magnet. — If a nail be heated red hot, a magnet will not attract it any more than it will a piece of brass; but when its temperature falls to about 1300°, it suddenly acquires nearly its full magnetic value, and will at once move promptly to the magnet if it be free to do so. This phenomenon means that the vigorous heat vibrations prevent the molecules from assuming any regularity of position until cohesion has become stronger.

The Lifting Power of a Magnet depends upon the strength of its field, and also upon how much of the field can be made to go through the iron which it lifts.

The strength of the field depends upon how large a proportion of the molecules are properly faced, and also upon how close the poles are to each other; that is, a U-magnet will hold up much more, proportional to its weight, than a straight bar will; and small magnets are generally stronger proportionally than larger ones. A good U-magnet should hold up three or four times its own weight, but small ones have been made that would hold up twenty-five times their own weight. These are values for permanent steel magnets. Electro-magnets can be made with sustaining power of a thousand pounds per square inch. It must be remembered that it is *ether pressure* that causes all the phenomena in a magnetic field, and hence the ether pressure may be as much as 1000 pounds per square inch.

Electro-Magnetism. — In order to detect the presence of an electric current in a wire and also its direction, the wire with the current is held over a magnetic needle, and the needle turns this way or that. One may now attend to a phenomenon not alluded to at first, namely, the needle moves *without touching the wire.* In a vacuum it moves still more readily, and one must conclude that there is some kind of a medium between the wire and the needle by which the needle is acted on, making it move. If the current be a strong one — several ampères — the needle will set itself nearly at right angles to the wire.

If a wire carrying a current be thrust through a piece of smooth, thick paper, or, better still, through a hole in a pane of glass, and filings of iron be scattered about the wire, and if the surface be gently tapped so as to permit the particles to arrange themselves, if there be any pressure upon them tending to move them, they will be seen to arrange in concentric circles about the wire (Fig. 47), showing there is a field about the wire very much as there is a field about a magnet. Let a wire with a current of several ampères be dipped into iron filings and they will adhere to it, forming a kind of coating to it a quarter of an inch thick; but on stopping the current, they will all fall off. Each bit of iron was made a magnet for the time, with north poles all facing one way, and so each one attracts and sticks to the next one in front of it, forming a circle about the wire. The wire does not attract them, but they attract each other, because they are magnets properly facing each other.

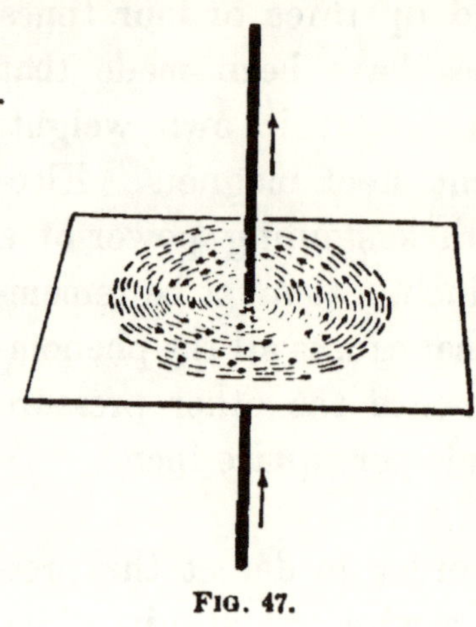

Fig. 47.

Now let the wire with the current be bent so as to make a single round loop three or four inches in diameter, and bring this loop near the needle, turning the loop so that first one side and then the other is near each pole. Each pole of the needle will be repelled by one *side* of the loop and attracted by the other, and the needle will set itself, if it be allowed, in the middle of the loop and at right angles to it; that is, it will be at right angles to every part of the loop.

By coiling the wire in this way, a greater length of it with its current can act upon the needle. If two turns be made, there will be twice as much wire and twice the current action upon the needle ; and ten turns will give ten times as much, and so on. If, then, a coil of wire be made of any shape, and a current of electricity be sent through it, one end or side of the coil will attract and the other will repel one end of the needle. That is to say, *a coil of wire with one or more turns, having an electric current in it, is a magnet*, with all the properties that belong to a permanent magnet, so long as the current lasts. It is the current alone that produces these effects, for it is the same whether the conductor be of copper, German silver, iron, or any other metal. Such a wire has a *field* extending to an indefinite distance from it in every direction at right angles to it, and the strength of this field varies inversely as the square of the distance from the wire. When the wire is made into a circle, the field is condensed inside it, and two, three, and more turns make it two, three, or more times stronger.

Iron is a much better conductor of magnetic stress than is the air. If a piece of iron be placed in the coils, the magnetic field of the wire is absorbed by it, and it becomes a magnet by *induction.* Its strength depends upon the strength of the current, that is, upon how many ampères the wire carries; also upon how close the wire is to the iron, and is greatest when the wire is wound upon the iron ; the strength of the magnet depends also upon how much wire is wound about it, that is, how many turns there are. The product of the current in ampères, multiplied by the number of turns of wire, is called *ampère turns.*

Such a magnet is called an *electro-magnet*, and as already stated, it can be made very much stronger than a permanent magnet. Its poles, too, can be reversed by reversing the direction of the current, which process reverses the position of the iron molecules. The rotation of the molecules brought about in this way results in friction among them in the iron, which becomes heated, and energy is thus wasted. This phenomenon is called *hysteresis.*

Electro-Magnetic Induction. — An electro-magnet is made by creating a magnetic field about a piece of iron. The magnetic field may come from another magnet already made, or from a current of electricity in a coil of wire about the iron. The action is a reversible one; that is, a magnetic field produced in a coil of wire will induce a current of electricity. This is what was shown on page 117, where was described the second method of generating electricity by moving a coil of wire in a magnetic field. If, instead of employing a permanent magnet as in that experiment, we should take an electro-magnet and wind it with a separate coil, so that the first coil could be connected with a battery or other source of electricity, while the second coil could be connected with a galvanometer, as in

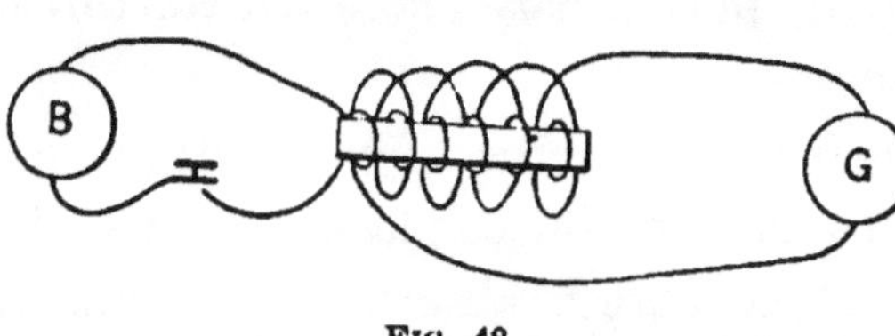

FIG. 48.

figure 48, where the inner coil represents the wire for an ordinary electro-magnet connected so that by pressing the key the current could be sent through it, the outer coil about the same rod leads to the galvanometer G. Every time the key closes the battery circuit the iron is made a magnet, a field is produced about it, and the second coil has a current *induced* in it which will show itself by the movement of the galvanometer needle. Holding the key down, the needle will come to rest at O, showing there is no current present in the second wire. On opening the first circuit, the magnet will lose its magnetism, the field will be destroyed, and the needle will again show a current in the second circuit, but in the *opposite direction* from the former one. This will happen as often as the first circuit is closed and opened.

If a key be so made as to reverse the direction of the current instead of breaking it in the first circuit, which is called the *primary circuit*, the current in the other, which is called the *secondary circuit*, will be much stronger. Such a magnet provided with two separate coils or circuits is called an *induction coil*. Of course it may be made of any shape or size, and have many uses, some of which will be mentioned further on in this book. Here it will be sufficient to point out the relation of the two coils to each other.

Suppose we have a bar of iron with two similar coils upon it; that is, both have the same number of turns of wire of the same size. This can be arranged by winding the wire on two spools, and putting one on each end of the magnet. Either of these may be used as a primary circuit. Suppose there be 100 turns of wire on each spool, and a battery be connected to the primary coil that shall give 10 volts' pressure in it, and a key for reversing the current be joined in the circuit; then every time the current in the primary circuit is reversed, there will be a pressure of 10 volts in the secondary coil. If there were 1000 turns in the secondary instead of 100, the pressure in it would be 100 volts; that is, the pressure in the secondary of an induction coil varies as the number of turns of wire in the coil; the size of the wire makes no difference in this particular. The pressure in the secondary circuit will be to the pressure in the primary circuit, as the number of turns of wire in it is to the number of turns in the primary; that is,

V in primary : V' in secondary : : t in primary : t' in secondary,

where V and V' are volts and t and t' are the number of turns of wire. This will be true only when the primary current is alternating in direction.

On page 149 it is pointed out that the strength of an electro-magnet depends upon the *ampère turns* of wire, and the same is true of the secondary wire.

Thus, if there be 100 turns and 1 ampère in the primary and 1000 turns in the secondary, the current in the latter will be such that the ampère turns in it equal the ampère turns in the primary. Since in the primary there are $100 \times 1 = 100$ ampère turns, there must be 100 ampère turns in the secondary; there are 1000 turns of wire in the secondary, the current in it will be $\frac{100}{1000}$ = one-tenth of an ampère. If the primary has 1000 turns and 1 ampère, the secondary with 100 turns will have 10 ampères.

As t in primary : t' in secondary : : c' in secondary : c in primary.

This is the principle involved in induction coils, and when employed thus it is evident that the voltage in the secondary is raised or lowered from that in the primary by simply varying the proportional number of turns of wire in it. As an induction coil transforms high voltage to low, or *vice versa*, it is called a *transformer.*

The energy in an electric circuit has been explained as being *equal to the product of the pressure in volts multiplied by the current in ampères.* This is equally true for electro-magnets and transformers. In primary, $ec = e'c'$ in secondary. For instance, let the voltage in the primary of a transformer be 1000, and the current 1 ampère, the energy $ec = 1000 \times 1 = 1000$ watts. The transformer hands over this *energy* from the primary circuit to the secondary; it does not increase it, it may diminish it by some losses. If the voltage of the secondary is reduced, say to 100 volts, still the number of watts is $e'c' = 1000$, and as $e' = 100$, c' must equal 10.

Heating Power of an Electric Current. — An electric current always heats the conductor through which it goes; how much it will be heated depends upon a number of factors. If 746 watts are equal to a horse-power or 550 foot-pounds of work per second, then $\frac{550}{746} = .75$ foot-pounds is the horse-power equivalent of one watt. As one heat unit equals 778 foot-pounds, 1 foot-pound is $\frac{1}{778}$ of a heat unit, and .75 of a foot-pound $.75 \times \frac{1}{778} = .00096$. This means that 1 watt, in 1 second, will heat a pound of water .00096°, and may be considered as the heat equivalent of a watt in a second.

From Ohm's Law we have $E = CR$. If this value of E be substituted for the E in EC, which represents electrical energy, we shall have $CR \times C = C^2R$ as an equivalent expression for electrical energy in terms of current and resistance. It shows that the ***energy or heating power is proportional to the square of the current***, and this must be remembered. Two ampères will heat 4 times as much as 1, and 10 ampères 100 times as much, and so on. If the resistance be doubled, the heat will be doubled, that is, the heating is proportional to the resistance. It might be supposed that if the current was maintained for 2 seconds, it would develop twice as much as it would in 1 second, and in t seconds t times as much. Putting all these factors together, the amount of heat developed is equal to $.00096 \times c^2rt$.

For instance, how much heat would be developed in 1 minute by a current of 10 ampères in a wire having 3 ohms resistance?

$.00096 \times 10^2 \times 3 \times 60 = 17.28$, or 1 pound of water would be heated 17.28° by it.

An apparatus suitable for studying the heating power is shown in the adjacent figure (49). It is a bottle for holding a known quantity of water, through the cork of which is fixed a coil of wire of known resistance, submerged in the water; a thermometer extends into the water, on which the rise in temperature may be observed during the passing of the current.

The temperature to which a given body will be raised by a definite amount of heat depends upon its specific heat (p. 92). As water has the highest specific heat, its temperature is changed the least by a given amount of heat. The specific heat of iron is .1138. The current of electricity that would heat a pound of water 17.28° in 1 minute would heat a pound of iron $\frac{17.28}{.1138} = 152°$. In any case, the heat required may be readily computed if the resistance, current, time, and specific heat be known, the formula being $\frac{.00096 \times c^2rt}{\text{spec. heat}}$.

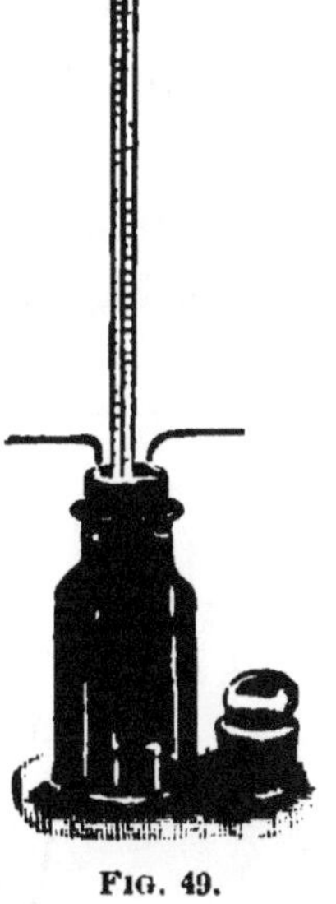

FIG. 49.

If the resistance be large and the current weak, the change in temperature may not be very noticeable. Telegraph lines are not much heated by the currents employed, but when the current is a hundred or a thousand ampères, the heating effect is ten thousand or a million times greater than when it is of one ampère, and metals and most refractory things may be fused.

In the electric-welding process the ends to be welded are pressed tightly together, while a current is sent through their junction strong enough to raise the parts to a welding heat. When they become cold, the junction is solid. The resistance is much greater at the place of contact than elsewhere, and therefore the heat is chiefly developed there.

The same principle of heating is applicable to cooking apparatus. A coil of German silver wire may be kept at any desirable temperature by a proper current of electricity. Ordinarily 8 or 10 ampères through a coil of 6 or 8 ohms' resistance is sufficient. This is not far from an electrical horse-power. $c^2r = 10^2 \times 8 = 800$ watts.

Another interesting method of electrical heating of metals has been discovered. If a current of 40 or 50 ampères be sent through a bucket of water made alkaline and a better conductor by adding some sodium carbonate to it, a high degree of heat is developed at the terminals dipping into the liquid. If one of these terminals be a large sheet of lead or of carbon, while the other be a rod of iron to be heated (Fig. 50), the latter may be fused under the water in a few seconds. The principle here is the same as the preceding, the conditions being a strong current and resistance.

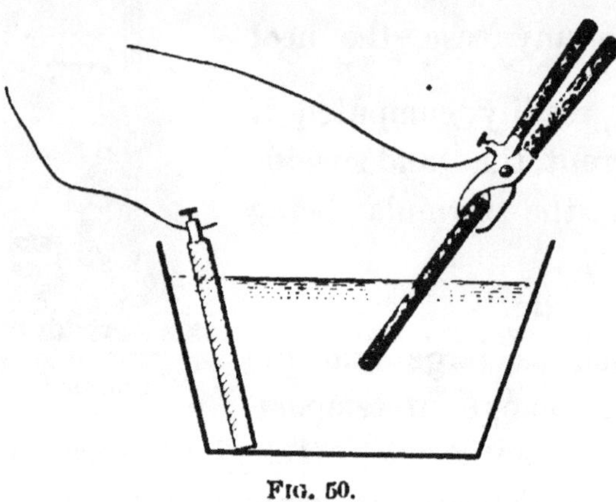

FIG. 50.

ELECTRIC LIGHTING.

The Arc Light. — When an electric circuit is broken, a spark may be seen at the place. If the pressure be 40 or 50 volts, the current is not broken when the metallic circuit is broken, and the ends of the circuit may be separated a distance that depends upon the voltage and the kind of material of the broken conductor. It is customary to use sticks of carbon, so fixed that the current keeps them separated about the eighth of an inch (Fig. 51), and raises their ends to a white heat; they then give a bright light. The energy is spent in the carbon tips. If the voltage be 45 and the current be 10, $EC = 45 \times 10 = 450$ watts; $\frac{450}{746} = .6$ of a horse-power. This electric energy is measured by connecting an ammeter in circuit with the arc lamp, and connecting the terminals of a voltmeter to the two carbons.

FIG. 51.

In Fig. 52, **L** is the lamp, **A** the ammeter, and **V** the voltmeter, **CC** the arc-light circuit.

All lamps in the circuit require the same number of volts. The voltage in an arc-light circuit depends upon the number of lamps lighted. The lamps are all in series, and the current is the same for all, and is generally 9 or 10 ampères. There are seldom more than 100 such lamps in one circuit, and as each one requires 45 volts, the dynamo must provide $45 \times 100 = 4500$ volts and 10 ampères. The *EC* of the circuit is $4500 \times 10 = 45000$ watts, $\frac{45000}{746} = 60 =$ horse-power.

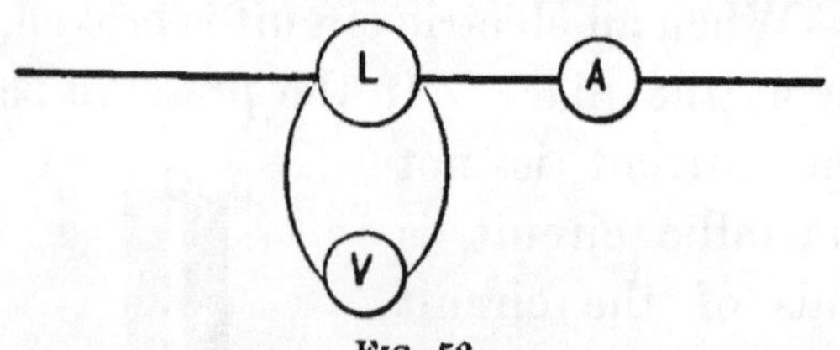

FIG. 52.

An arc lamp using 450 watts gives a light equal to about 800 candles. *Search lights* use a current of 100 and sometimes 200 ampères at 45 volts, and the beam reflected from a concave mirror behind the lamp is concentrated into a dense beam, reckoned at millions of candles.

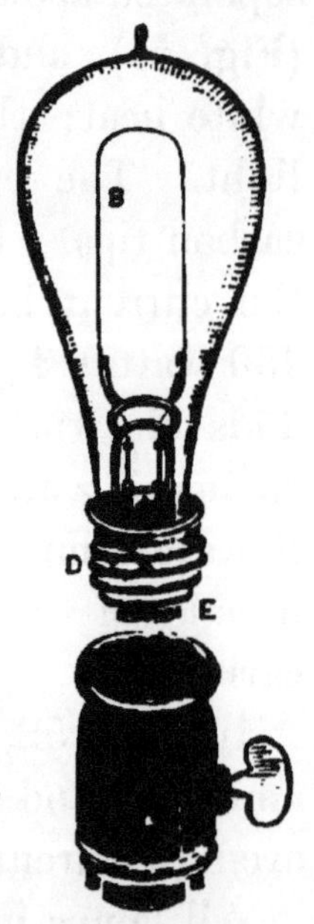

FIG. 53.

The Incandescent Light. — This is the name given to the light from a thread or filament of carbon, which is heated white-hot by a current of electricity passing through it. When carbon is thus heated in the air, it not only gives out light, but it is consumed; that is, its surface molecules combine with oxygen of the air, forming gas. To prevent this action the filament is enclosed in a glass bulb from which the air has been exhausted. The lamp is familiar enough to all (Fig. 53). The light from it is directly due to the high temperature which the electric current produces, not to electricity itself; so the stronger the current, the higher the temperature. Most such lamps have filaments of such a size that, with a current of .5 or .6 of an ampère, a linear inch gives the light of 2 or 3 candles. The longer the filament, the more light. A sixteen-candle-power lamp has a filament about 8 inches long. Such lamps are adapted to particular electric pressure, as 50 volts or 110 volts.

A 110-volt lamp using a current of .6 of an ampère uses $110 \times .6 = 66$ watts of electrical energy. If the lamp now gives the light of 16 candles, each candle requires $\frac{66}{16} = 4.1$ watts. By increasing the current to .7 of an ampère, the light may be increased to 25 candles, and the amount of energy *per candle* is lessened, for $112 \times .7 = 78.4$ watts, and $\frac{78.4}{25} = 3.1$ watts. By increasing the voltage to 115, the current may rise to as much as an ampère, and the light to 100 candles; but when this is done, the filament is soon destroyed.

If one will recall what is happening when a body is heated, — that its molecules are violently beating against each other, and at the surface are being evaporated with a rapidity which depends upon the temperature, — he will understand what happens to an incandescent lamp filament when the glass becomes blackened and the filament broken by the vigor of the molecular vibrations.

Unlike arc lamps, which are arranged in series, incandescent lamps are arranged like the steps on a ladder; this is called multiple connection: thus (Fig. 54), **a** and **b** are wires from the dynamo, and have an electric pressure between them of 50 or 110 volts. Any lamp having its terminals connected to these 2 wires will have a current through it, the strength of which depends upon the resistance of the lamp. Thus, if the voltage be 50, and the lamp has a resistance of 100 ohms, the current will be $\frac{50}{100} = .5$ of an ampère. The watts would be $50 \times .5 = 25$, and, at 4 watts per candle, such a lamp would give about 6 candle-power.

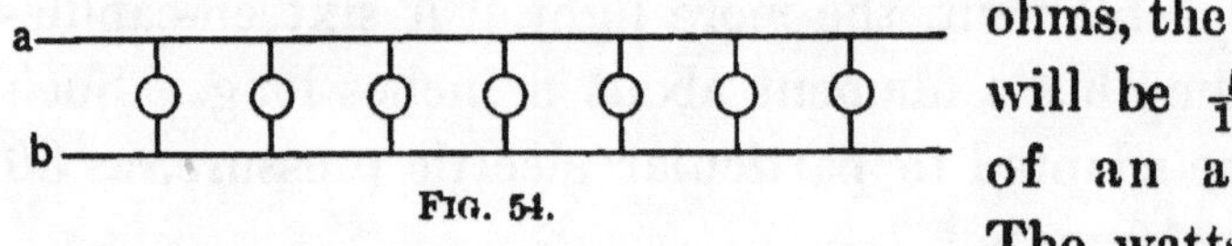

FIG. 54.

One may now see the relation between horse-power and electric lights. The electric horse-power is equal to 746 watts. If an incandescent lamp requires 4 watts per candle, a horse-power will give $\frac{746}{4} = 186$ candle-power, which may be distributed among as many or as few lamps as one pleases. If each lamp gives 16, there may be $\frac{186}{16} = 11$ lamps per horse-power; if the lamps give but 10, there may be 18 + such lamps to the horse-power.

If an arc light consumes 450 watts = .6 horse-power, and gives light of 800 candle-power, a whole horse-power would give 1333 candles, seven times as much as with the incandescent system. And if 746 watts give 1333 candles, each candle requires $\frac{746}{1333}$ of a watt, or only little more than half a watt per candle.

The expenditure of energy in an incandescent lamp may be measured in precisely the same way as in an arc lamp.

The Dynamo. — If a wire be put between the poles of a U-shaped magnet, it will be in the strongest part of the magnetic field. If it be moved towards the bend in the magnet, an electric pressure will be developed in it, the direction of which is indicated by the arrowhead upon the wire (Fig. 55); and if the ends be connected to a galvanometer of proper delicacy, its needle will move. When the wire is moved back a current is induced in the opposite direction.

If two wires be thus moved at the same time, each will have a current induced in it in the same direction,

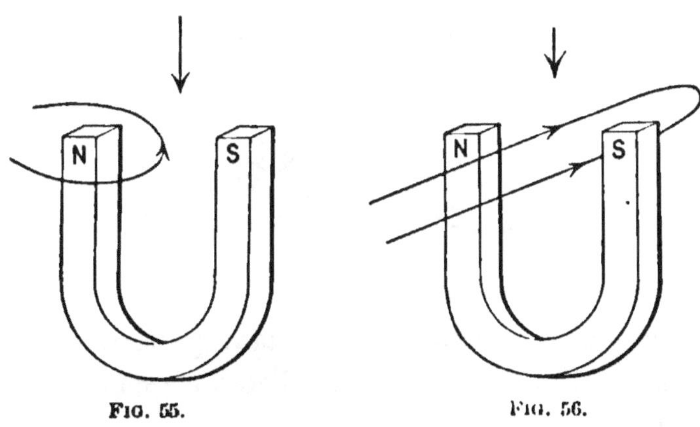

Fig. 55. Fig. 56.

and so on for as many wires as may be thus moved. If the two are parts of the same wire, as indicated in Fig. 56, the two pressures will balance each other, and no current at all will flow. If, however, the loop of wire be fixed to rotate so that its mechanical motions will be opposite, — one going up through the field while the other is going down, as in Fig. 57, — a current will flow around the wire as the arrows point. For one-half of the revolution the current will be in one direction, and the other half it will be in the opposite; it is therefore called an *alternating current*. If the two ends of the wires be connected to the two halves of a metal cylinder which are insulated from each other upon the axis, and against which some springs make contact (Fig. 58), the current may be led off by wires to any place where it is wanted, and it will be continuously in the same direction.

This device for directing the current is called a *commutator* — an exceedingly important part of electrical machines. Such a current is called a *constant* or *direct current.*

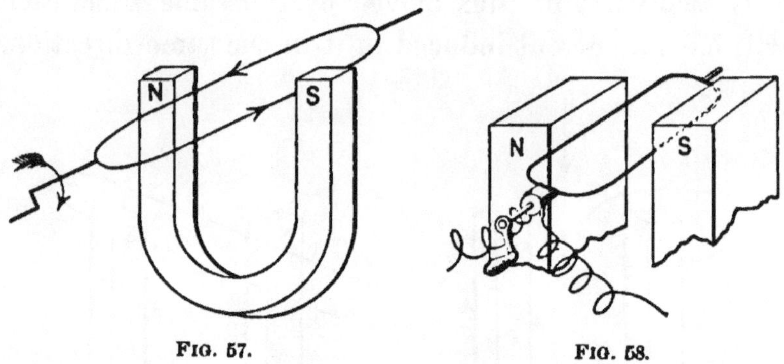

FIG. 57. FIG. 58.

By adding more turns of wire to rotate together in the same field, the pressure is increased proportionally; it depends also upon the rate of rotation. The pressure is also dependent upon the strength of the magnetic field; the stronger the field, the higher the voltage. The strength of the field may be greatly increased by putting iron between the poles of the magnet, as it is a better conductor of magnetism; hence the wire is wound upon iron, generally in cylinder or ring form, and the poles of the magnet are curved so as to be as close as possible to the rotating parts. The large magnet **N S**

(Fig. 59), that provides the magnetic field, is called the *field magnet*, and the rotating part of iron with its coils of wire is called the *armature*. Together they constitute a dynamo.

The first machines of this kind were made with field magnets of steel permanently magnetized. Now they are made of soft iron with coils of wire about them,

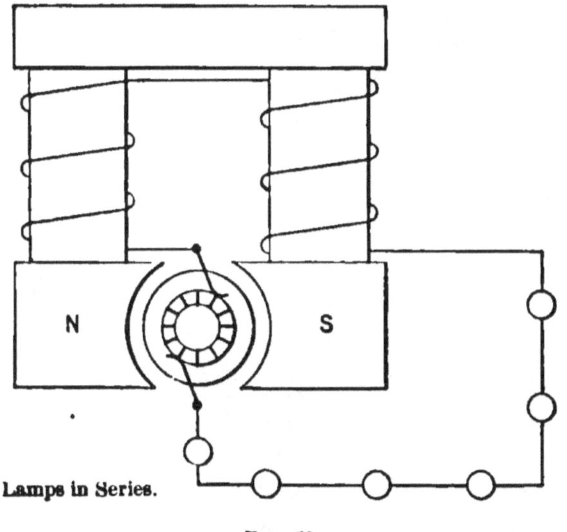

FIG. 59.

that is, they are electro-magnets, and the current generated in the armature is made to pass through the field-magnet coils (Fig. 59); hence the stronger the current, the stronger the magnet. Such an arrangement is called a *series dynamo*, and is the one generally employed for arc lighting.

In some dynamos only a portion of the current generated goes through the field-magnet coils; the remainder goes through the main circuit, to be used for commercial purposes. Such a machine is called a *shunt dynamo* (Fig. 60), and is the kind employed in incandescent lighting. When run at uniform speed, it is capable

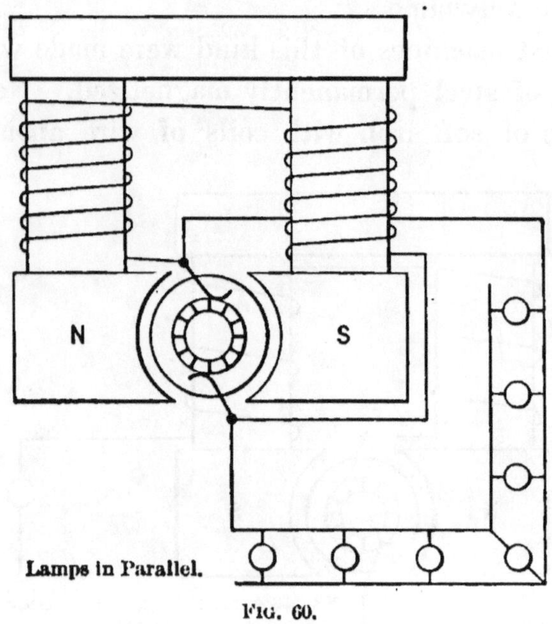

Fig. 60.

of giving a uniform electric pressure, as 50 volts or 110 volts throughout its main circuit; the lamps are connected in parallel circuits to the larger wires, called *mains*.

Either of these machines may be made of any size, and their field magnets may be shaped in any convenient way without changing their electrical qualities.

A dynamo is a machine for transforming mechanical energy into electrical energy. Its efficiency may be as high as 97%. That is, if a steam-engine of 100 horse-power be made to drive a dynamo, and if there be no loss, the electrical energy, EC, would be $746 \times 100 =$

74,600 watts. That would be 100% efficiency. As a matter of fact it would be no more than 74,600 × .97 = 72,362 watts, and might be less on account of heating, friction, and so on. As it is, however, it is one of the most perfect machines man has yet made. The dynamo may be made of any size and of any horse-power, up to 5000 or more, so as to furnish electrical energy of any pressure and current.

The Electric Motor. — This in structure is simply a dynamo. When a current of electricity from some other source is made to traverse its armature coils, the magnetic reaction turns the armature, and it may be

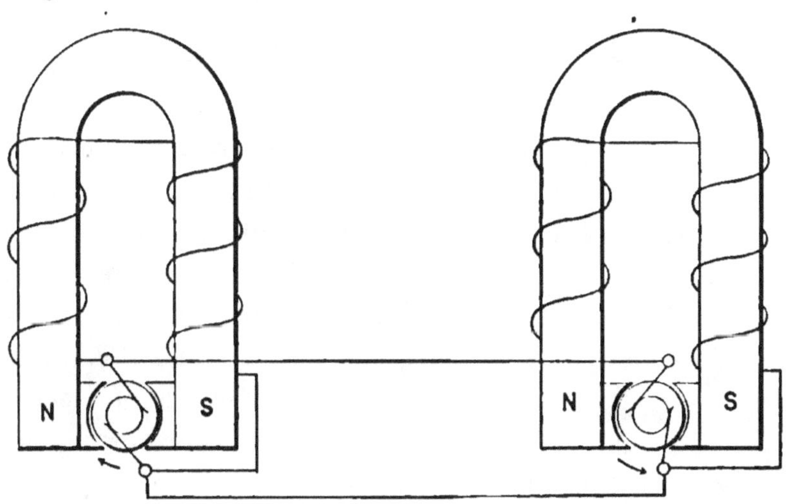

FIG. 61.

made to do work. When thus employed, the same machine is called a *motor* (Fig. 61).

The function of a motor is to transform electrical energy into mechanical energy, and its efficiency is also high.

It should be remembered that electrical energy is always derived by transformation from some other kind of energy, and can do no more mechanical or other kind of work than an equal amount of energy of any other kind can do.

THE TELEGRAPH.

I. The Electro-Magnetic.—The ability a current of electricity has to make a magnet of a piece of iron, which can attract to itself another piece of iron and stop attracting it when the current stops, is employed for making signals at a distance. In Fig. 62 **e** is an electro-magnet, **a** the iron armature upon an arm, **s** a spring to keep the armature up from the magnet and against the stop above. A corresponding stop below prevents the armature from quite touching the end of the magnet. In Fig. 63 **B** is a battery or other source of electricity; **k** is a key for closing the circuit. As often as the key closes the circuit a current goes through the coil, making **e** a magnet which pulls down the armature **a**. This action is so quick that the stroke of the armature can be plainly heard, taking place whenever the key **k** is worked by the hand. A series of strokes made faster or slower can be produced, and these have been constructed into a telegraphic

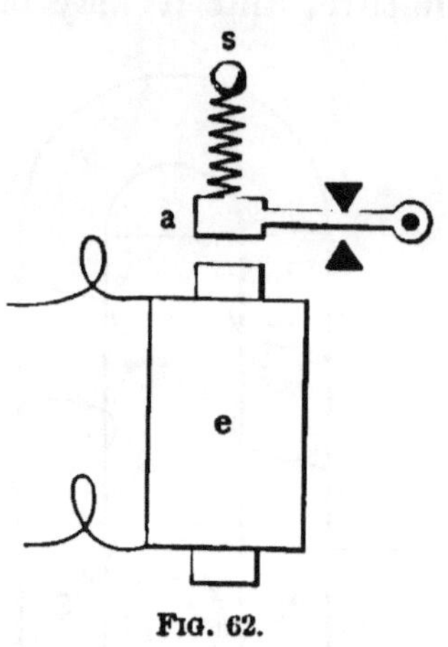

FIG. 62.

alphabet, the letters of which can be recognized by the ear, the length of time the key is held down being told by the sound of the up-strokes. Thus a short and a

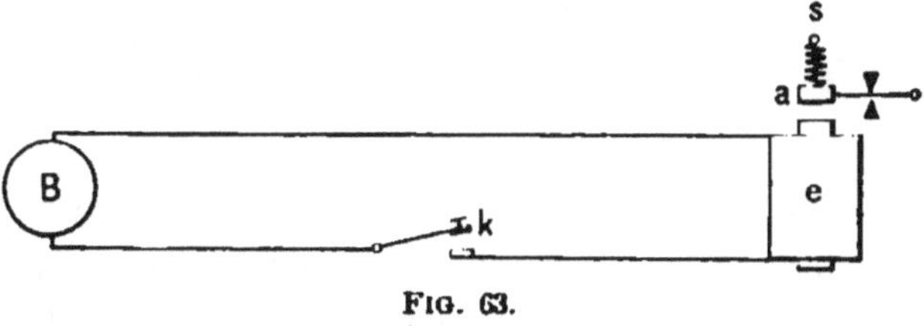

FIG. 63.

long hold give the letter *a*, a long and three short ones the letter *b*, and so on. This is called *reading by sound*, and the magnet adapted for this work is called a *sounder* (Fig. 64). How long the line is does not much matter, for it is the current which is needed for the electro-magnet, and the increased resistance of a longer line may be met by using higher pressure. Thus, suppose a sounder requires the tenth of an ampere to work it properly, and has a resistance of 5 ohms. If the line be a mile long with the resistance of 10 ohms, and the pressure is 5 volts, the current will be $\frac{5}{5+10} = .33$ of an ampère. If the line be made ten miles long, the resistance would be 105 ohms; and in order to get .1 of an ampère, there will need to be $105 \times .1 = 10.5$ volts. For long lines it is customary to use magnets that will

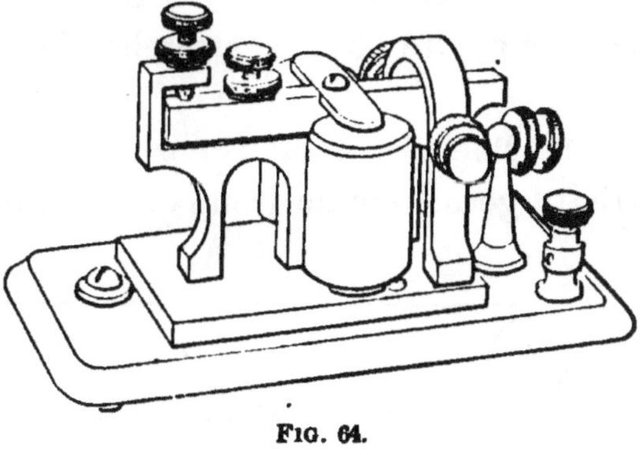

FIG. 64.

work promptly with currents no stronger than .01 of an ampere, and often these are so made that in working they open and close another electric circuit which extends still farther away. The latter device is called a *relay*. The resistance thus overcome may be as much as 150 ohms.

A continuous current implies a continuous conducting circuit which must not be broken ; and a conducting circuit means a continuous line of conductor from one terminal of the battery or other source of electricity to the other terminal. It is not necessary for this conductor to be wire; the earth, when damp, acts as a conductor if the ends of the wires are buried in it. A buried plate of copper or iron at the end of each wire seems to decrease the resistance of the circuit, yet each such buried plate adds about 100 ohms to the circuit; 200 ohms would be no more than the resistance of 15 or 20 miles of wire. So if the line be a long one, only one wire need be stretched between stations.

II. The Chemical Telegraph. — If a current of electricity goes from an iron point through a piece of paper

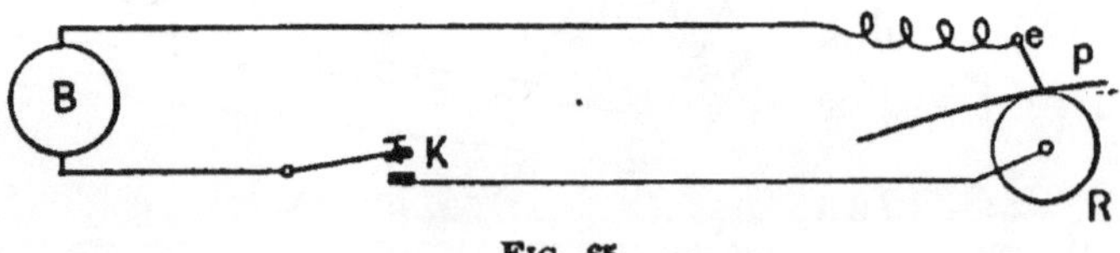

FIG. 65.

moistened with the ferrocyanide of potassium, it makes a blue mark. If the paper be drawn along, it will make a blue line.

Suppose **R** represents a metallic roller on which the strip of paper **p** is drawn, while the iron finger **e** touches upon it, the finger being connected to the battery **B**. When the key **K** closes the circuit, a blue mark will be made where **e** touches the paper. If **K** be worked, say once a second, while **p** is being drawn along in the direction of the arrow, there will be a series of blue marks made. If the key be worked as for the electro-magnetic alphabet, there will be a series of dots and dashes corresponding to the letters, which may be read by those acquainted with them. As the action is purely a chemical one, this telegraph is called the *chemical telegraph*. It makes no sound, is much more sensitive than the former one described, and can be worked at a higher rate of speed.

Mechanical Transmitters are made by punching holes and slots corresponding to the letters in the tele-

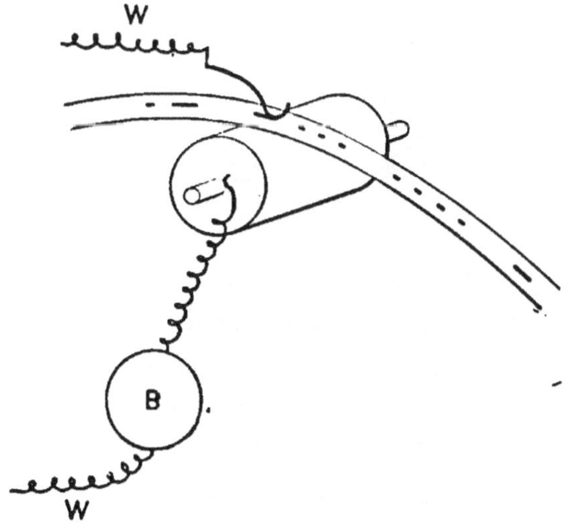

FIG. 66.

graphic alphabet in long strips of paper, and drawing this over a metallic roller. A metallic finger presses upon the paper, and falls into the holes as the strip is pulled along, and so completes the electric circuit for a longer or a shorter time. The chemical telegraphic receiver makes corresponding blue marks on a strip of paper; this is done at the rate of four or five hundred words a minute. With the ordinary hand-key, thirty or forty words a minute is a good working rate.

THE SPEAKING TELEPHONE.

The Magnetic Telephone.—If one holds a sheet of paper in front of his mouth and sings or talks, he can feel the jar of the vibrations that the air waves produce on the paper. It moves to and fro as many times a second as there are air waves per second. If he does the same with a thin piece of sheet iron, the same effect

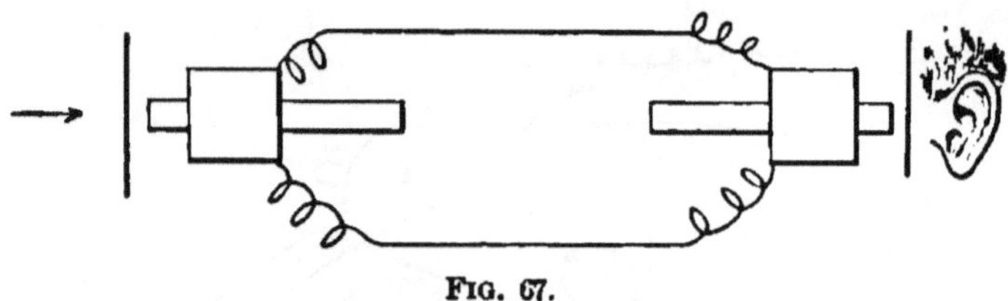

Fig. 67.

follows, only not quite so appreciably, as the iron has more mass to move. If the iron is brought up near the pole of a magnet about which is a coil of wire, the inductive action between the magnet and the iron causes electric currents to go on in the wire this way or that, as the plate moves towards or away from the pole. It is a kind of dynamo in which the armature is moved by sound waves. If connecting wires lead to a distant similar instrument, the currents will cause the attraction between the magnet pole and its iron armature to be now stronger and now weaker, and will

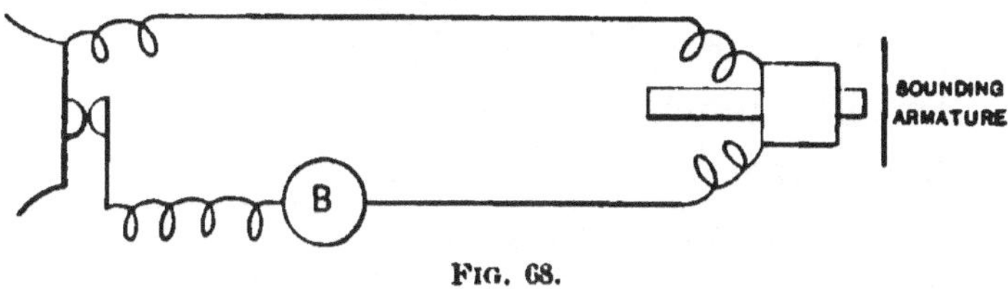

FIG. 68.

make the plate vibrate in the same way as the plate in the first instrument vibrated; and if the first vibrations were produced by speech at the arrow (Fig. 67), an ear at the other instrument will hear what is said, the latter instrument acting as a motor.

If the plate spoken to be made to press more or less hard upon a knob of hard carbon through which a

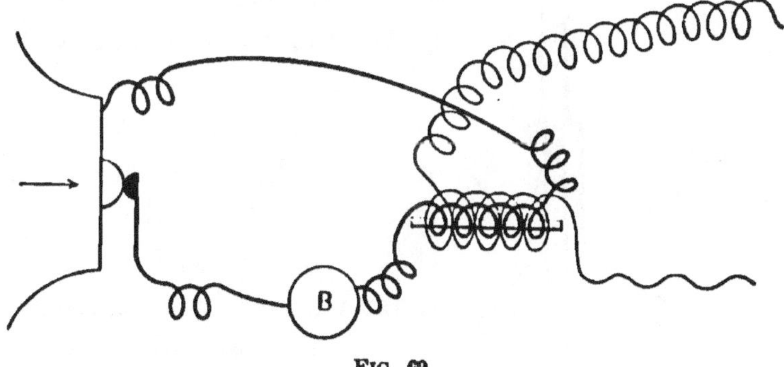

FIG. 69.

current from a battery is going (Fig. 68), the varying pressure, when sound waves act on the plate, will cause the current to vary in a corresponding way, and the same result is produced at the listening instrument or *receiver*, as it is commonly called. The instrument spoken to is called the *transmitter.*

Generally the transmitter is combined with a small induction coil (Fig. 69), so the battery current is in a very short circuit, the secondary wire going to the receiver. This arrangement is much more efficient than the former one, and by its means one may talk to another a thousand miles away.

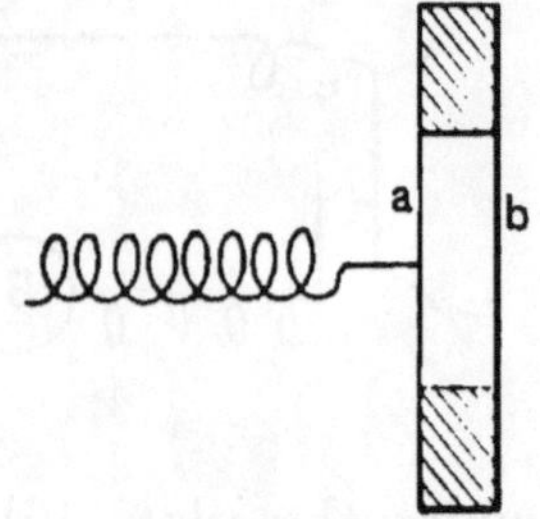

FIG. 70.

The Static Telephone. — There are several other methods by which sounds of any kind may be reproduced by electricity. One of them is by means of electrical attraction as distinguished from magnetic attraction. Thus, if a wire from any source of electricity be connected to a metallic plate **a** (Fig. 70) by some method, another metallic plate **b** near it will be attracted by it whenever the plate **a** is electrically charged. If it be prevented from making contact with **a** by separating the two by a non-conducting ring, the middle of the plate **b** will bulge toward **a**, and its own elasticity will restore it to flatness when the attraction stops. Hence a series of electric charges from a source, such as a transmitter

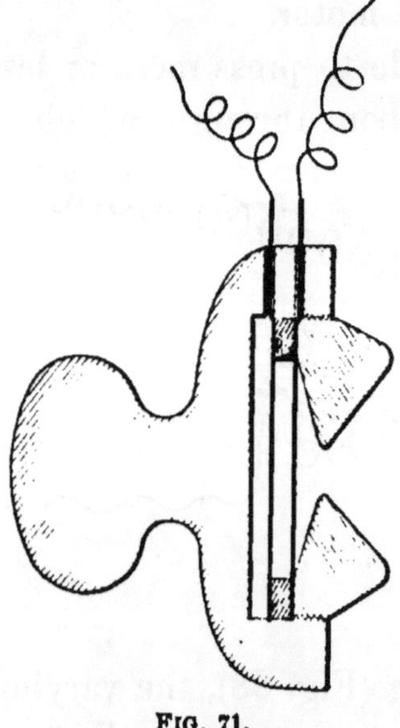
FIG. 71.

with induction coil for high pressure, will make the plate **b** move in a corresponding way, and speech may

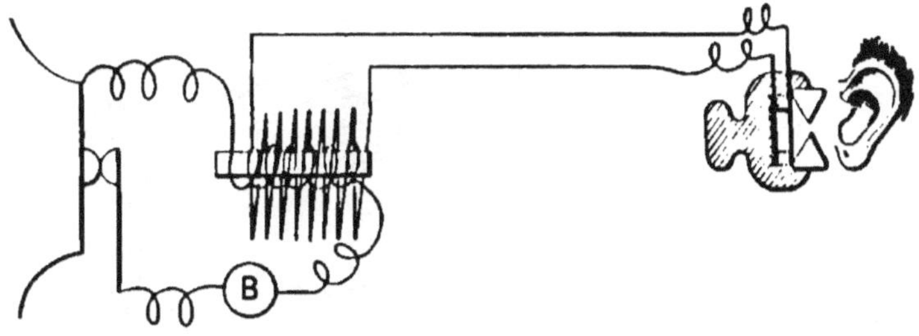

FIG. 72.

be heard by listening at **b**. The whole arrangement is shown in diagrams (Figs. 71 and 72).

TELEGRAPHING WITHOUT WIRES.

By Conduction. — When the two ends of an electric circuit are buried in the earth, the current through the earth spreads out in every direction, but chiefly goes toward the opposite terminal. Thus, **a** and **b** (Fig. 73) represent the ground terminals of a circuit, the stations being connected by a wire through the air as in the ordinary telegraph. The lines that radiate from **a** go to **b**. If **a** and **b** be 4 or 5 miles apart, the earth current may be as much as 2 miles broad at **c d**, and a wire 100 feet long stretched at **d e** along those lines, and having its ends stuck into the ground, will have that part of the current travel through it. If an ordinary telephone receiver be connected in this wire, any changes made in the circuit **a b** may be heard in the short circuit at **d**. Telegraphic signals may therefore be sent to **d** from either **a** or **b** without any wire directly connecting it with **d**. If **a** and **b** were on one side of a river, communication could be made with **d** on the

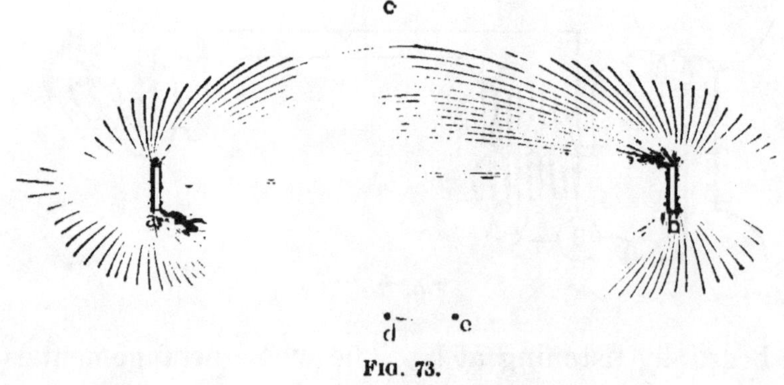

FIG. 73.

other side. The further apart **a** and **b** can be placed, and the stronger the current employed, the greater may be the distance of **d** from the direct wire connecting the two stations.

By Induction. — When the wires of two different circuits are carried on the same poles, as is the common way, the telegraphic or other currents in one induce corresponding currents in the other, so that if the second one have a receiving telephone in it, all the signals may be heard and speech transferred from one circuit to the other. Telegraphic signals have been heard when the parallel lines were as much as 3 miles apart. The explanation lies in the fact that every current in a wire sets up a magnetic field about it in every direction, as explained on page 148, and this extends to an indefinite and much greater distance than was imagined before the telephone was employed as a detector. The facts make it necessary to conceive electric waves radiating from a wire carrying vibratory currents, as they do from a vibrating magnet, to immense distances, probably limitless, only growing presently too weak to be detected by present methods. When there is a complete metallic circuit, and the wires are near together like those on telegraph poles, the induction effect is much greater between them, and consequently less strong at a distance. This arrangement serves to protect telephone lines from induction troubles. If the telephone line is itself double, or a metallic circuit with the wires near together, the induction effect on one is neutralized by an equal effect on the other.

Electro-Chemical Work. — It has already been pointed out that electricity can decompose water, and also that these elements are not set free at the same place, but one at each of the terminals. Let the diagram (Fig. 74) represent a tank containing acidu-

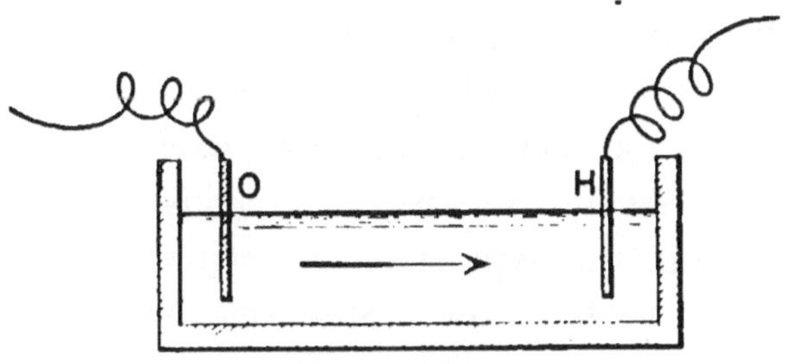

FIG. 74.

lated water into which two platinum strips dip, and through which a current is made to go in the direction shown by the arrow. Oxygen gas will be set free at the terminal where the current enters the liquid, and hydrogen at the terminal from which it leaves it. By providing suitable tubes (Fig. 75), these gases may be collected and the quantity measured. It is found that the volume of hydrogen is twice that of oxygen, which corresponds with the formula for water, H_2O, signifying twice as many atoms of hydrogen as of oxygen in water. If in place of water some chemical solution, such as sodium sulphate, be tested in the same way, another curious chemical result follows. The salt is decomposed, the sodium is all set free at one terminal, and the acid at the other. To observe this effect, it is best to color the liquid with purple cabbage solution (purple cabbage boiled in water). The current will change the color to green on the sodium side, and red on the acid side.

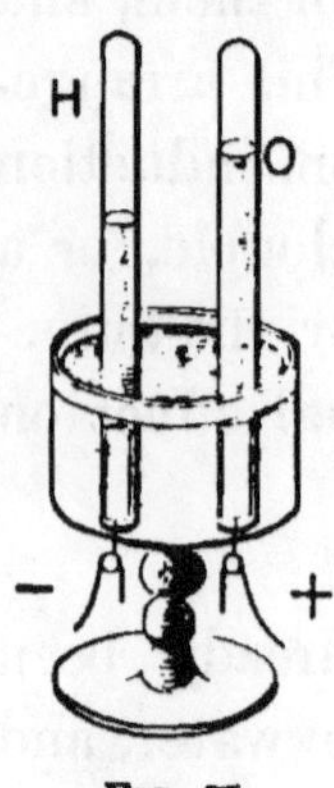

FIG. 75.

In this case it is the molecule $NaOH \cdot SO_3$ which is decomposed into NaOH and SO_3. The NaOH is an alkali which turns vegetable colors to blue or green, while SO_3 is an acid and turns the same to red. The molecules of the sodium sulphate exchange partners along the whole line of the current just as the water molecules do in the experiment described on page 127. The process is rather slow, unless one has a current of as much as an ampère. The water will not be decomposed appreciably unless the pressure be as much as a volt and a half; but the molecule of $NaOH \cdot SO_3$ will not require so much, that is to say, a single galvanic cell will suffice, though the work goes on slowly with but one. Some chemical industries now employ electricity on an extensive scale. Sodium carbonate is thus manufactured by decomposing common salt in a tank like that in Fig. 74. The sodium is set free at the **H** terminal, and forms with water sodium hydrate, which is alkaline. Carbonic acid gas is driven in a stream through the alkaline solution, and chemical combination results in a solution of sodium carbonate. When this is evaporated, the solid of commerce is left.

Chlorate of potash is also manufactured by decomposing in a similar way potassium chloride in a solution of potassium hydrate KHO. Chlorine and oxygen are set free at the **H** terminal, and these combine with the caustic potash to form the chlorate. Thousands of tons of each of these products are made in this way every year.

If a solution of copper, silver, or other metalic salt be employed, a current will decompose it and deposit the metal upon the terminal from which the current leaves the solution. It is in this way that *electro-plating* is done. The object to be plated is immersed in a solution of the salt of the metal with which it is to be covered, and a plate of the same metal is placed in the solution. The current goes from the latter to the former through the liquid, and the dissolved metal is deposited in a uniform coating upon the object. Special preparations have to be made, as the details for plating with silver are different from those best adapted to nickel or copper plating.

The amount of this chemical work done depends upon the strength of the current in ampères, as well as the length of time employed, for evidently twice as much can be done in two minutes as in one minute.

The weight of any chemical element which may be set free from its chemical combination by a current of one ampère in one second — that is, by one coulomb of electricity — has been very accurately determined.

As a coulomb will decompose .001422 grains of water, the weight of hydrogen set free will be $\frac{.001422}{9} =$ .000158 grains.

The weight of other elements set free or deposited as in plating is equal to this weight of hydrogen, .000158, multiplied by their $\frac{\text{atomic weight}}{\text{valency}^{1}}$.

Thus, for

$$\text{Oxygen, } .000158 \times \frac{16}{2} = .001264.$$

$$\text{Zinc, } \text{“} \times \frac{65.2}{2} = .005150.$$

$$\text{Copper, } \text{“} \times \frac{63.5}{2} = .005016.$$

$$\text{Silver, } \text{“} \times \frac{108}{1} = .01706.$$

The meaning of these numbers is simply that a current of one ampère will deposit in one second .01706 grains of silver upon a metallic surface exposed to it. They also show that there is a quantitative relation between the current and the chemical work it will do, and the quantity of metal deposited depends upon the quantity of electricity in coulombs, as well as upon the atomic weight of the metal.

[1] Valency is a chemical term meaning the number of hydrogen atoms the given element is chemically equal to.

Secondary Batteries. — If the wires connected to the apparatus (Fig. 74) be connected to a galvanometer after the apparatus has been decomposing water, the needle will show a current of electricity which results from the recombination of the hydrogen and oxygen. Its pressure will be 1.5 volts. The current will last no longer than the time required for the chemical work due to the slight amount of gas at the platinum terminals. It is to be noted that the terminals are both of the same metal, and under ordinary conditions no current would be furnished. If two plates of lead be employed instead of platinum, both immersed in a solution of dilute sulphuric acid, and a current of electricity be sent through them and the liquid, one of them will be coated with the peroxide of lead, PbO_2. After this chemical action has taken place the cell with its plates is a galvanic battery, and can give a current of electricity until the peroxide is decomposed. The cell will have an electric pressure of 2 volts. After it has yielded up its current it may be again charged by a current forming the peroxide as before; then it will be ready to give a current again. This process may be repeated an indefinite number of times. Such a cell is called a *secondary battery*, and is in very common use to-day. It is customary at the beginning to apply to the two plates of such a cell an artificial coat of oxide of lead, in the form of a paste, and then send the current through as before; this shortens the process of formation. The action of the charging current is not to charge the plates with electricity, but to change their chemical condition. When properly prepared, the two plates are as unlike chemically as are zinc and carbon, and give a current of electricity for precisely the same reasons so long as the chemical work can go on. When the plates are large, having several square feet surface, they may receive a relatively large amount of electric energy and change it into chemical energy. They may yield a current of twenty or thirty ampères for several hours at a pressure of 2 volts. The resistance of such a cell is small — a few thousandths of an ohm. To yield a horse-power, EC must equal 746 watts. If the current be, say, 25 ampères, then there must be $\frac{746}{25} = 30$ volts nearly, and this requires 15 cells. As each cell weighs as much as forty pounds, it is seen that a battery to yield a horse-power would weigh about six hundred pounds. This is one reason why such cells have not been employed more extensively; another is that they are not very durable.

High Electric-Pressure Phenomena.—It has already been said that an electric arc consists of a current of electricity between two separate carbons. It remains to be pointed out that in electric arc circuits the carbons have to touch each other in order to start the current. The current will not jump from one terminal to the other. Nevertheless, there is a real tension at the carbon points, as may be seen by attaching a voltmeter to them; but ten thousand volts' pressure is only sufficient to give a spark the tenth of an inch long. Induction coils are made capable of giving pressures of hundreds of thousands of volts, and consequently much longer jumping distances. A spark an inch long implies a voltage of 75,000, two inches twice that, and so on. Sparks have been artificially produced five feet long; they therefore represented between four or five million volts.

Induction coils for producing sparks several inches long (Fig. 76) have a secondary coil of a great many turns of fine wire, often several miles long, wound upon a spool, which may be taken off of the primary coil and magnet. A strong current of ten or twenty ampères in the primary coil may be interrupted by hand or by some automatic arrangement. There are terminals to the secondary which allow adjustment at different distances. The appearance of the spark is the same as that of lightning. Its duration is so brief that a swiftly moving object lighted up solely by the spark appears at rest. The spark may be made much brighter by connecting the two surfaces of a Leyden jar to the opposite terminals of the secondary coil; the spark will be shortened to an inch or less and will be very noisy.

FIG. 76.

If the air be partially exhausted from a glass tube three or four feet long, the spark may be sent through it, indicating that a partial vacuum is a better conductor than air at ordinary pressure. Yet if the air-pressure be greatly reduced, the flash will not pass at all through the tube; it will jump three feet in the air rather than the tenth of an inch in the exhausted tube. This shows that for the *conduction* of electricity matter is necessary.

Tubes called Geissler's and Crookes' tubes are made in great variety for showing various electrical effects

I.

FIG. 77.

due to high pressure. In Geissler's the rarefaction is not very great, but the tubes contain traces of gases of different kinds; each kind has some tint different from others, and may in this way be studied with the spectroscope.

In Crookes' tubes the variety of phenomena is very great, ranging, indeed, through a large part of physics.

The mill (Fig. 77) has a wheel which is made to whirl by the impact of molecules upon it, as in the radiometer.

Tube (Fig. 78) has a platinum disk in the middle of it, and the impact of molecules raises the temperature of the disk to a red heat.

Tube (Fig. 79) shows that the molecules go in straight

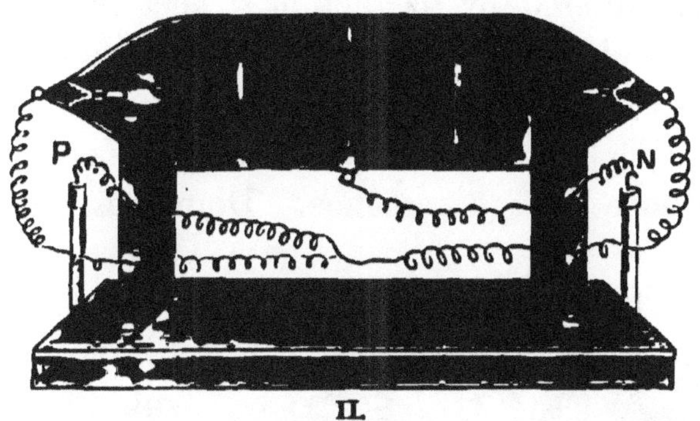

II.

FIG. 78.

lines, and also that glass is fluorescent when bombarded in this way. A metallic Greek cross enclosed in it may be raised so as to shield the back of the glass from the

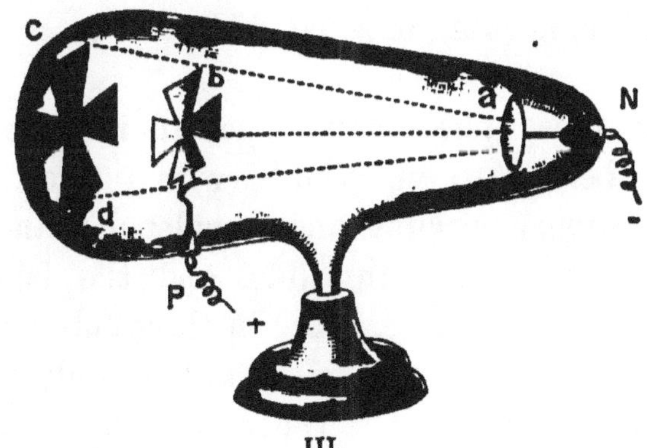

III.

FIG. 79.

molecular shower, and an image of the cross may be seen upon the back of the glass. When the glass has been lighted a short time in this way, if the cross be dropped so as to expose the surface which had been protected, and then the discharges be repeated, the outline of the cross will be seen brighter than where the glass had suffered by the action. This indicates that the glass molecules become fatigued, and need time to recover their original sensitiveness.

Tube (Fig. 80) shows how such electrified molecules may produce phosphorescence. Diamonds, rubies, and many common crystals glow with brilliancy when subject to such action.

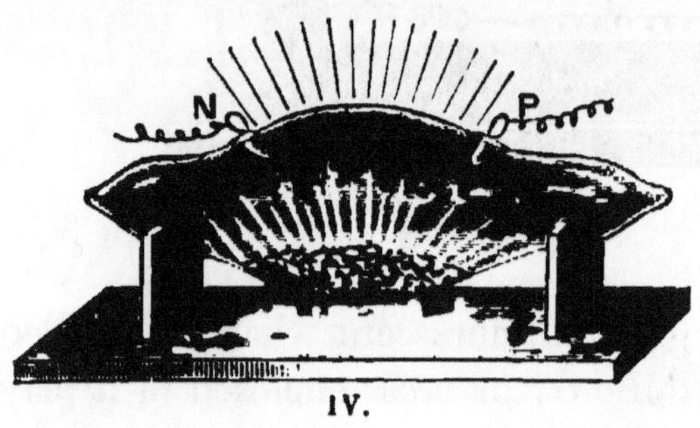

FIG. 80.

A magnet held near one of these tubes shows that an electrified stream of molecules may be deflected by a magnetic field.

All of these phenomena have an importance that does not belong to many otherwise interesting facts. Under ordinary circumstances molecular and atomic phenomena are quite invisible, and the behavior of molecules has to be inferred. In these tubes the molecules are luminous, and one can see not only what they do, but how they do it. When the wind makes the windmill go round, one understands that it is due to the impact of the air as it strikes the sails. The mill in the tube goes round for the same reason. The metallic terminals within the tubes are highly energized by the electric pressure to which they are subjected. The gas molecules rebound from them, not so much because they are electrified as because they are vigorously pushed away, as is the case with the radiometer (p. 250). Their velocity is greatly increased, and they act like a wind blowing upon the vanes of the wheel. At the same time the molecules are made incandescent, and their direction may be seen.

When the bullet strikes the target, the temperature of both is raised, for mechanical motion has been changed into molecular vibratory motion. In like manner, in tube 78, the impact of a great number of molecules, each having high velocity, act individually, like bullets, and heat the platinum plate, which is the target against which they strike; and it is heated red-hot if the energy be great enough. It is not electricity that heats it; it is mechanical impact.

The law of motion explained on page 34 is that a body in translatory motion will continue on *in a straight line* until some other body acts upon it. In tube 79 this is shown to be true of molecules, for the sharply defined shadow of the cross upon the back of the tube is due to that space being shielded from impact, and behind the cross the space is dark, while the rest is light. The fatigue of the molecules already alluded to is only another example of the same condition brought about by overwork in ordinary matter. By overwork is meant making the matter transform energy at a rate so high as to change its own qualities. An overworked machine breaks down. Overworked molecules lose their elasticity.

The luminous effects in both gas and glass indicate high rates of vibration, chiefly atomic, and represent impacts at rates which compare with the vibratory rates of the atoms themselves; that is, are more or less sympathetic (p. 281), and therefore do not so hastily die out, as do forced vibrations. This phenomenon is called *phosphorescence* or *fluorescence*.

The magnetic field, which may be shown with iron filings, may be seen to affect the stream of shining molecules in any of the tubes, making them move in curved paths. This is an ether effect, and shows that a mass of matter cannot go in a straight line through ether which is in a magnetic stress.

STATIC ELECTRICITY.

There is another class of phenomena due to electricity which are not ordinarily shown by electrified bodies, on account of the weak effects when the pressure is no more than a few hundred volts. When this reaches several thousands, phenomena of attraction are observed. A glass rod rubbed with a piece of silk or flannel becomes electrified; that is, has electric pressure developed upon it, and light bodies in its neighborhood are attracted to it. In like manner two strips of dry brown paper drawn between the thumb and forefinger become electrified similarly, and repel each other. If the hand be put between the loose ends, they will both approach it, and will recede from each other on its removal.

Two balls a half-inch in diameter, made of the pith of a sunflower stalk, may be hung on silk threads 6 or

8 inches long, and serve for several interesting experiments with electricity developed in this way. If the rubbed glass rod be brought near to them when they are suspended as in Fig. 81, they will both advance to meet it and touch it; but directly they will leave it, and then it will be found that they are repelled very strongly from the same rod. Let a stick of sealing-wax be rubbed with the same silk, and brought near to the pith balls that the rod repels; they now will be attracted by the stick. Here are evidences of two kinds of electrification. They are called *positive* and *negative.* The glass rod gives what is called positive, the wax negative.

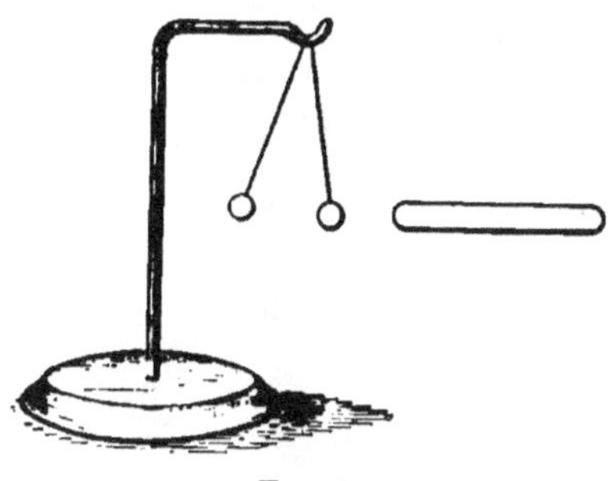
FIG. 81.

The Electroscope (Fig. 82) is a device for identifying these two electrical conditions. It consists of a pair of gold-leaf strips **p**, 3 or 4 inches long, connected to a metallic stem going through a glass tumbler to a knob or shelf on top at **b**. The tumbler rests upon a metallic support, and from this 2 strips of tinfoil **c** and **c′** pasted upon the inside of the glass reach into the tumbler so the gold leaves can reach them. When a sufficiently electrified body is brought near to the electrometer, the gold leaves diverge. On removal they collapse. If the finger be touched to the top knob while the leaves diverge, and *be removed before* the *exciting body* is, then, removing the latter, the leaves will remain apart. They are themselves electrified, and are very delicate indicators of the presence of electrified bodies.

Now electrify the glass rod as before, and slowly bring it near; observe whether the leaves diverge still further or collapse. Whichever way it be, the excited stick of wax will affect them in the opposite way. As the excited glass is called positive, the movement of the leaves indicates how it is affected by a positive

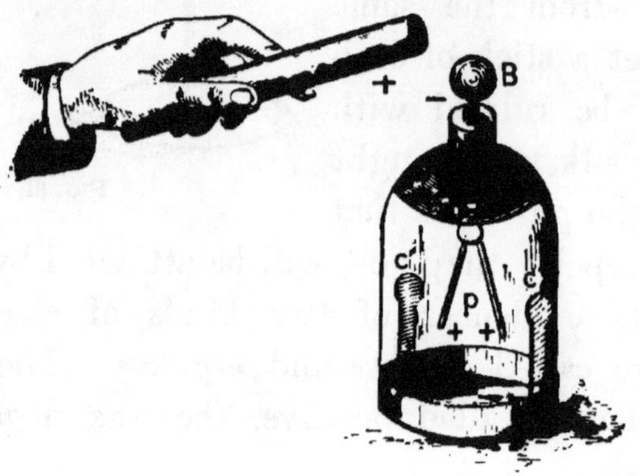

FIG. 82.

charge. Other bodies, such as books, wood, hard rubber, crockery, etc., may be rubbed and tested in like manner. Another experiment is to rub the glass with the silk as before, and test the silk. It will be found negatively electrified.

When two bodies are rubbed together, and one of them is electrified, the other one is oppositely electrified.

A small sheet of celluloid makes an excellent object for observing such phenomena with. By rubbing it with either silk or woolen, it will become so highly electrified that sparks an inch long may be drawn from it.

Whether a given body be positively or negatively electrified depends upon what it is rubbed with. A piece of hard rubber rubbed with woolen becomes negative; if rubbed on silver it becomes positive and the silver negative. All substances may be electrified in some degree by friction. Those which exhibit it under ordinary conditions are those called non-conductors; that is, if they be electrified in any way, the electricity is not conducted away. Take a metallic rod of any kind, and hold it in a silk or woolen cloth wrapped around one end of it; then rub the free part with a piece of silk or cat-skin, and test it, as in the other cases, with the electroscope. It will be found electrified. If it be held in the hand without the wrapping, — which is called insulation, — the electricity developed will be conducted away as fast as it appears. Such phenomena indicate that all kinds of substances can be electrified in the same way, and that such as cannot conduct electricity from molecule to molecule remain electrified. The molecules of the surface of the glass that has been rubbed are all electrified, and remain so because conduction does not take place.

The Electric Field. — The phenomenon of magnetic attraction is explained as due to the magnetic field which exists in the space about a magnet. Attractions and repulsions take place because the ether pressure is greater or less upon a body that can be acted on by it. In like manner an electrified body has an *electric field*, within which all substances whatever are attracted or repelled. There is no such choice as there is with magnetic substances. When a body is electrified by rubbing or in any other way, it reacts upon the ether so as to produce a stress in it, and in two different ways, glass giving what is called positive, and wax negative, characters. Each one produces a field of such a sort that another body brought into it becomes electrified in the opposite sense; that is, a positively electrified body *induces* negative electrification upon that side of another body which is next it, and positive upon the remote side. With small bodies it is not so easy to show this as it is with larger ones; thus, let **A** (Fig. 83) be a body that has been electrified in any convenient way, by letting 2 or 3 electric sparks fall upon it. Then if one end of an insulated cylinder with rounded ends, **B**, be brought near to it, but not allowed to touch it, the cylinder will be found to be positively electrified at one end, and negatively at the other. To test this, take a thin disk of metal the size of a quarter of a dollar, fastened to the end of a stick of wax for a handle; touch the electrified body **A** with this disk, which is called a *proof plane;* then to discover whether it has positive or negative electricity, hold it near the excited electrometer. Then touch the disk with the finger to discharge it, after which it may be touched to either end of the body **B**, and tested in like manner. This

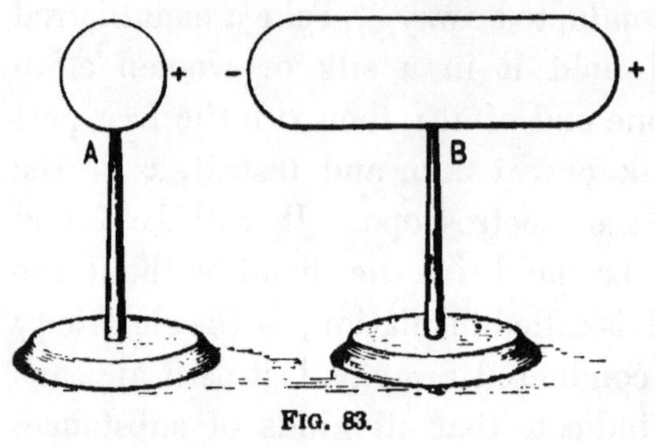

FIG. 83.

will remind one of the magnet which induces opposite polarity in a piece of iron on the end next the pole, and the same polarity on the further end of it. A mechanical idea of what takes place may be got by considering the small ring **A** (Fig. 84) to be attached by spiral springs on every side to a larger body around it. If **A** be *twisted* to the right or left, it will pull on all the springs, which will all act to bring it into its normal position as soon as it is free to move. If twisted to the right, the springs will pull to the left; and if it be twisted to the left, they will pull to the right. If one of these directions be called a positive, the other may be called a negative, stress, — a right-handed or a left-handed twist.

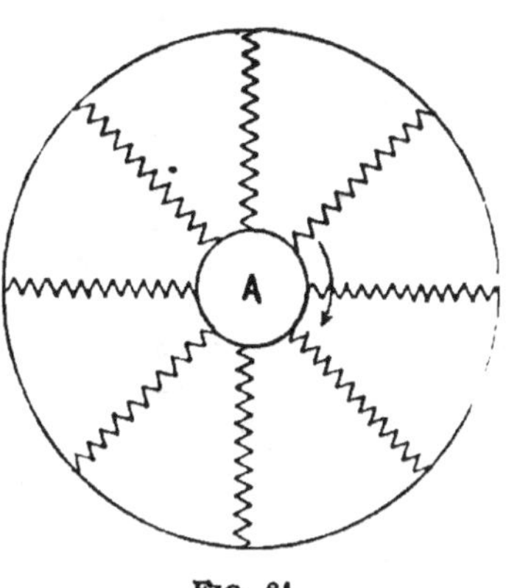

FIG. 84.

If there were two such bodies **A** and **B** (Fig. 85), each connected by similar springs about it, and also to each other, then it is plain that if **A** were twisted in either direction, it would tend to twist **B** in the opposite direction. Also, if there were another one in contact with **B** on its right-hand side, it would be rotated in the same direction as **A**. If **A** were right-handed, **B** would be left-handed, and the next in line right-handed. This is only a mechanical analogy to help one to a conception of what takes place between molecules in an electric field. The ether that is about all bodies holds their molecules in certain relative positions, and when they are made to assume some new position, it tends to pull them back; that is, the ether is put into a state of stress by abnormal positions of molecules, and another

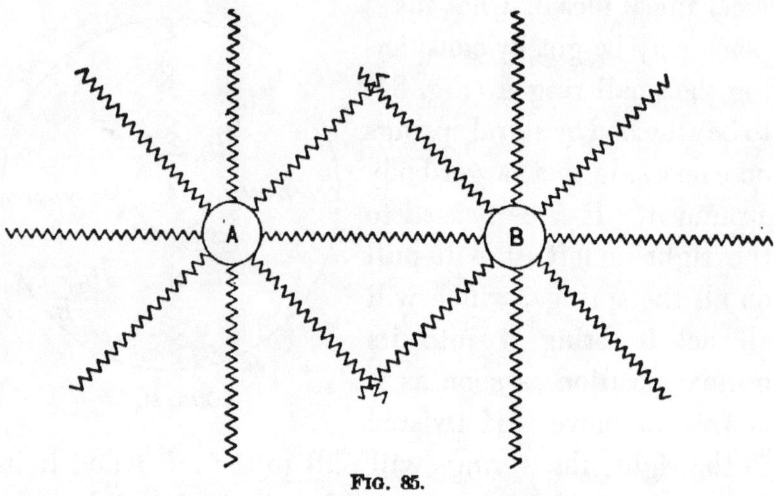

FIG. 85.

molecule brought into that space is twisted into an opposite position, which constitute what is called the positive and negative conditions, and the action is called *electrical induction.*

STATIC ELECTRICAL MACHINES.

I. **Frictional.** — By rotating a piece of glass against which silk or other substance presses, it will be subject to continuous friction, and may in that way be electrified. By presenting metallic points near to the surface that has been thus electrified, electricity may be collected by them, and the metallic part to which the metallic points are fixed may become strongly electrified. Such a glass plate-machine (Fig. 86) works well in dry weather, but not in damp, for moisture collects on the glass and other parts of it. The electricity is of very

FIG. 86.

high voltage; the moisture is a sufficiently good conductor for electricity, so it leaks away and cannot be accumulated.

II. Inductive. — There are several kinds of machines which depend upon inductive action; that is, some part is electrified at first, and its inductive action upon the glass plate or plates which may be rotated serves to keep them electrified; and by means of the metallic points, electricity is collected as in the frictional machines. The action of such machines is complicated. Some of them need to be separately excited, others are self-excited, and most of them are less affected by dampness than the frictional machines. With such a machine (Fig. 87) in good order, having plates 20 inches in diameter, one may get sparks 4 or 5 inches

FIG. 87.

long, and many interesting experiments may be made with such high-pressure electricity.

STORAGE.

The Leyden Jar. — This consists of a glass jar (Fig. 88) coated inside and out, nearly to the mouth of the jar, with tinfoil pasted on it. Through the cork is a metallic conductor with a chain touching the inner foil, and ending at the top with a ball. If such a jar be held in the hand while the knob is touched to one of the knobs of the electric machine while the latter is worked, the jar will become *charged*. It will contain electricity in what is called its static form, and will retain it for a long time if pains be taken to prevent its discharge. Such a charged jar may be discharged by connecting the inner and outer surface by any conductor. The knuckle will do, but it is likely to give one a shock, so a device called a *discharger* is employed; it consists of metallic arms with a glass handle (Fig. 89). One may touch with safety either the knob *alone* or the outer surface *alone* of the jar; so it is entirely safe to hold it in the hand while charging it.

FIG. 88.

The charge resides upon the glass and not upon the metallic coating. This is shown by making the jar in separable parts: thus a glass goblet (Fig. 90) fitting into a tin cup that serves for the outer coating, a smaller tin cup fitting into the glass for the inner coating. A glass tube extends from the bottom of this inner jar to the knob on the end, to enable one to take it out without discharging it. When all are in place it may be charged by holding the metallic outer cup in the hand, and holding the knob from the inner cup to the electrifying machine until the jar is heard to hiss. Then, *after* setting it upon the table, remove the inner jar **D** by taking hold of the glass tube, after which it may be handled all over with safety; then the glass **B** may be removed and set by itself; the outer cup **C** will then remain, and this, too, may be handled like any tin cup. If they be put together again as at **A**, the whole may be discharged with a snapping spark with the discharger as at first.

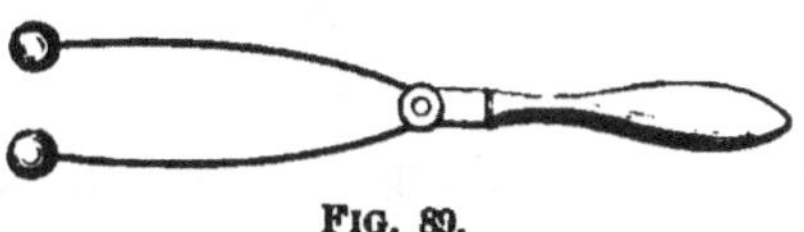

FIG. 89.

Now let the same jar be charged and taken apart as before. With the electrometer and proof plane (p. 190), test the two surfaces of the glass; they will be found to be oppositely electrified. If the hand be put into the glass without touching it, a crackling sound will be heard, and a sensation as if the hand were in a spider's web will be felt.

These are most interesting facts, as they have directly to do with electrical theory. The two sides of the glass, being oppositely electrified, are in an opposite state of strain. The molecules are twisted into new positions, and this condition is slowly conducted through from one side to the other, so it is not possible to wholly discharge the glass at once. Charge it in the common way, let it stand for five minutes, then discharge it. Let it remain for a minute or two, and another spark may be got from it. This shows that electrification is an affair of the glass molecules, and takes time for their action. If a wooden twig be twisted half-way round and then freed, it will fly back *most* of the way to its normal position at once; but the original position will not be reached for an appreciable time. If it be highly elastic, it will vibrate for a short time to and fro, twisting in each direction alternately. This is precisely what happens in the discharge of the jar; the electric current is not simply

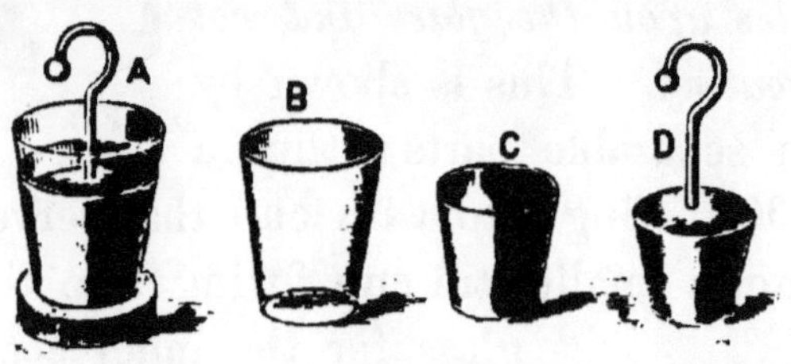

FIG. 90.

a spark in one direction, but an alternating current, the period depending upon the size of the jar. The latter, in discharging, acts like an elastic body, giving currents in both directions, and setting up ether waves having a length that depends upon the number of vibrations of the coating of the jar, just as the length of sound waves in air depends upon the number of vibrations of the body that produces them. For a gallon jar this number is hundreds of millions per second; so the wave-length may be a few feet long, for ether waves move at the rate of 186,000 miles a second, no matter how they originate or what their wave-length. Every charged body is to be considered as having its molecules in a state of strain, and every spark of discharge as relieving the *strain, and producing vibrations like elastic bodies.*

Metallic Points discharge electricity into the air without a spark.

A spherical body like a cannon ball or a bullet may be electrified, and retain its charge for a long time if it

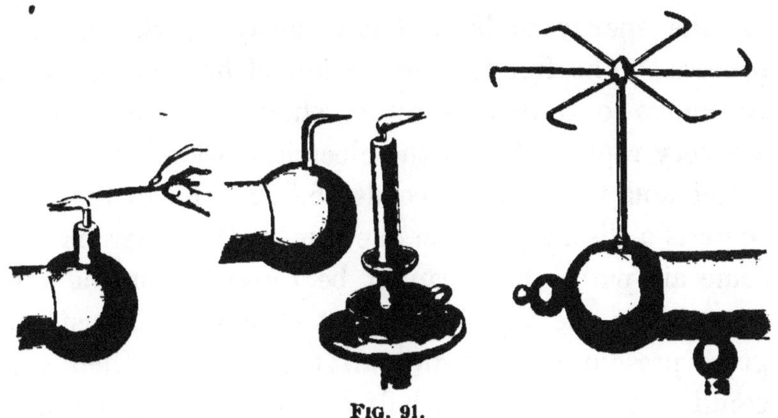

Fig. 91.

be perfectly insulated, so that the electricity cannot creep away on the support; but if it be provided with a point, it will be discharged almost instantly into the air. In like manner, if a point like that of a darning-needle be presented to a charged body (Fig. 91), it will discharge itself through the needle without the noisy spark which otherwise would be produced.

There is a mechanical reaction between the air and the point from which electricity is escaping, as is shown by mounting a number of pointed wires so they may rotate (Fig. 91). It will also blow upon a candle flame, and a strong discharge may extinguish the flame.

All of the experiments which may be performed with the induction coil (p. 181) may be shown with electricity developed with static machines. Indeed, there is no difference except in the degree and manner of generation.

Lightning is a phenomenon of electricity of exceedingly high pressure or voltage. It originates in clouds when condensation is taking place at a rate too high for the energy to be radiated away by the ordinary process. The frequent formation of hail as an accompaniment to thunder showers shows that the changes are very rapid. When the electric tension between the cloud and the earth exceeds 75,000 volts to the inch, there is a discharge from the one to the other through some air path which cannot be foreseen, for the temperature of the air, the amount of moisture, dust, and gases present are varying all the time. When such clouds are high, that is, half a mile or more, the lightning discharges go from one cloud to another; when less than that distance, the discharge is more frequently into the earth, generally by the way of a tree having roots that spread near a water course, or a damp locality that acts as a conductor from the cloud to the mass of the earth. The form of a flash of lightning is not zigzag, as has so often been pictured, but is rather a wavy line (Fig. 92), as is the spark seen from an induction coil. Its course is partly determined by the presence of dust particles along its line. There is no evidence that it has momentum.

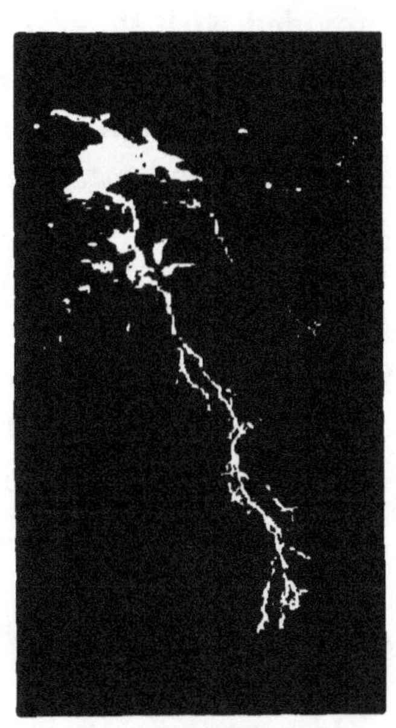

FIG. 92.

Flashes between clouds are of very short duration, generally less than the hundred-thousandth of a second; but some discharges to the earth last a second or more, and represent a great amount of energy, as is seen by the destruction they sometimes cause. It has been shown that some discharges are as great as 250 coulombs.

Protection from Lightning. — As trees are liable to be struck by lightning, they are unsafe places for shelter in a thunder shower. It is wiser to take a drenching than to seek the protection of trees. For the protection of houses it has long been the custom to attach metallic rods to the outside, with branches raised above the tops of chimneys and other higher points. The lower ends are carried into a well, moist earth, or connected to metallic water or gas-pipes, so as to convey any accidental discharge into the earth. Such rods are useful, but many buildings have been struck when provided with them in the most approved manner. It is rare that any one is seriously injured in a house thus defended, though sometimes the house itself is injured. Lightning and other high voltage alternating currents are easily interrupted in ordinary conductors and find new paths. They seldom keep entirely to the wire or rod provided for them. It is now deemed safer to employ flat straps of iron or copper stretched along the ridge pole and down the roof angles and each corner of the house to the ground, so as to have a kind of coarse net over the house, the lower end being carried into the ground as before. Similar straps should be carried up the outer parts of the chimneys, and fastened around them at the top.

Iron gas-pipe an inch in diameter is as good as anything for such purposes, and it has the advantage of being cheap and readily adjustable to any length. Platinum-tipped forks for the top of the rods are of no special value, and insulators are useless.

Conclusion as to the Nature of Electricity.—It is plain that whenever electricity is generated, some mechanical, chemical, or heat phenomena are its antecedents, that is, movements of masses of matter of some degree of magnitude. If they are movements of large masses of matter that may be seen, we call the phenomena *mechanical;* if they are too minute to be seen, we call them *molecular* phenomena; but the presence of matter, and some of its motions are indispensable. Also, when electricity does work of any kind, — turning a motor, decomposing molecules, heating a mass of matter, or giving light, — it is giving motion of one kind or another to the matter it acts upon, or it produces a pressure in it when it cannot move it, as when one presses with his hand against the wall. The amount of matter is not changed in any degree by any process. The only change taking place in any phenomenon is a change of motion of one kind or another, mechanical or molecular in matter, and stress or vibrations in the ether, and these motions are exchangeable with each other under proper conditions.

There is no known phenomenon in the ether which does not depend upon the prior action of matter to produce it. It follows that electricity is a phenomenon of molecules in which energy is in a different form from its common mechanical or heat form. The mechanical is known as translatory, the heat as vibratory; and the probability is very strong that electricity represents rotary motions among molecules and atoms.

QUESTIONS.

1. If a wire has a resistance of 10 ohms to the mile, what will be the resistance of 250 miles of the same kind of wire?

2. If a pound of No. 10 copper wire has a resistance of .033 of an ohm, what will be the resistance of 115 pounds of it?

3. A thousand feet of No. 25 copper wire has a resistance of 33 ohms; what will be the resistance of a mile of it?

4. If 1000 feet of No. 10 wire has a resistance of 1 ohm, what will be the resistance of the same length of wire which has twice the diameter?

5. What will be the resistance of the same length if the diameter be one-half?

6. What will be the resistance of iron wire in each of the above cases if copper be six times better than iron as a conductor?

7. No. 40 copper wire has a resistance of 1 ohm to the foot; what is the resistance per foot of iron, platinum, and carbon filament, if of the same diameter as the copper?

8. What will be the electro-motive force of 10 ammonium chloride battery cells if they are connected in series?

9. If a cell have an electro-motive force of 2 volts, and a resistance of 1 ohm, what current will it give through a wire so short and large that its resistance is too small to take account of?

10. What current will the cell give through a wire with 9 ohms' resistance?

11. A cell having an E. M. F. of 1.4 volts will give what current through the circuit of an electro-magnet having a resistance of 5 ohms, its own resistance being .5 of an ohm?

12. A battery cell is giving a current of 1 ampère when the resistance is known to be 2 ohms; what is the E. M. F. of the cell?

13. What is the resistance of a circuit when a battery with 1.8 volts gives a current of 3 ampères?

14. A telegraph sounder requires .1 of an ampère to work it properly; if a battery of 4 cells, each with 1.5 volts, be in series in the circuit, what must be the resistance of the circuit that it may be worked?

15. A telegraph wire 200 miles long has a resistance of 10 ohms to the mile; with a galvanic battery having an E. M. F. of 150 volts and an internal resistance of 100 ohms, how much current will be in the circuit if the grounds at the ends have 100 ohms each? Suppose there be 5 relays in the circuit, each with 150 ohms' resistance; how much current will now traverse the circuit?

16. An incandescent electric lamp is lighted by a current of 1 ampère, and the pressure is 100 volts; what is the resistance of the lamp?

17. If there be fifty such lamps arranged in multiple order in a circuit, how many ampères must be sent into the wires that lead to the lamps?

18. How many watts are spent in a lamp taking 50 volts and 1.2 ampères?

19. How many watts are there to 10 electrical horse-power?

20. How many watts are spent in an incandescent electric light circuit, with 110 volts and 100 lamps, each taking .6 of an ampère? What electrical horse-power is spent in the lamps?

21. If a battery cell having 2 volts' pressure furnishes a current of 2 ampères, what part of a horse-power does it yield?

22. How many foot-pounds of work can the same cell do in 1 second? In an hour?

23. What current must be furnished by a battery of 25 two-volt cells in order to produce a horse-power?

24. If a carriage requires 2 horse-power to drive it, and its motor takes 20 ampères, how many two-volt cells will be needed?

25. An arc-light circuit has 90 lamps in series in it. Each lamp requires 45 volts and 5 ampères to run it; how many electrical horse-power are spent in them?

26. What will be the resistance of each lamp on the above supposition?

CHAPTER X.

ETHER WAVES.

Origin of. — In the chapter on heat it is explained how heat originates from mechanical sources such as percussion, friction, and condensation, from chemical sources such as combustion, and from electrical currents; also how heat is transferred by the mechanical processes of conduction and convection from one body to another. There is still another way by which heat energy may be transferred from one body to another without the bodies being in contact, that is, without conduction.

If a cannon ball were heated to a red heat and hung by a wire from the ceiling, it would lose its temperature slowly, and one in its presence could feel the warmth from it upon his hands and face. A little attention will show that this is not due to the heated air that is all about it, for air currents will not be felt except over it, and a glass put between the hand and the hot ball will stop air currents, but not the sensation. A hot body will cool almost as freely in the most perfect vacuum that can be made as it will in free air, as is shown by the incandescent electric lamp which has a red-hot filament of carbon in a vacuum. When the current is stopped, the filament becomes black and cools almost instantly. Conduction is out of the question, and the energy represented by the high temperature finds another way of escape.

Again, that which we call light, whether from fire or sun or stars, gets to us in some way from bodies at all distances from us. Hold the hand in sunlight and warmth may be felt; yet we are confident that between the sun and the earth there is a space of 93,000,000 miles in which there is a more perfect vacuum than one can possibly produce in any artificial way. Energy gets to us across this space void of matter.

On page 138 the omnipresent ether is mentioned, and the properties of magnets and of electric currents

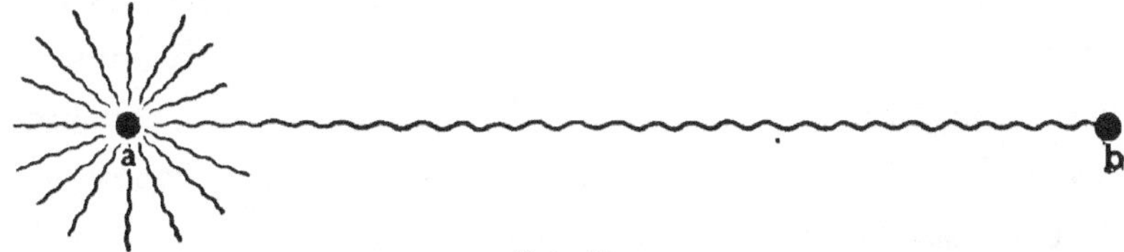

FIG. 93.

are explained as due to its action. Here again it must be considered as the agency by which heat energy is transferred from one body to another.

The atoms and molecules of all bodies are extremely minute (p. 7) and elastic (p. 16). When they possess much energy and collide with each other vigorously, they vibrate like bells or other sounding bodies, and thereby set up waves in the ether which travel in every direction, as in Fig. 93; and these waves continue to travel on in straight lines until they come to another mass of matter, as at **b**, which may stop them or turn them in some new direction. If the body **b** stops these waves, itself becomes heated by them, and its molecules vibrate like **a**, that is, the temperature of **b** rises. The waves thus produced in the ether, by vibrating molecules, are called *ether waves*, and the energy they represent is sometimes called *radiant energy*, because it is radiated from a hot or vibrating body.

Consider now what has mechanically happened as illustrated in Fig. 93, where **a** represents an atom or molecule of ordinary matter in a state of vibration; that is, is heated, has some temperature. It is surrounded with ether, and every vibration does work on the ether, producing a disturbance in it which assumes the form of a wave; and a series of vibrations cause a series of waves which follow each other. What was at first a vibratory motion of matter has now become a wave motion in the ether; the energy which at first was called heat energy has now become *radiant energy*. There has been a transference of it from one body to another, and a transformation from one kind to another; hence it is no longer proper to speak of the energy of the ether waves as heat energy; its only name now is radiant energy. When these waves fall upon other molecules, as at **b**, the energy is changed back to its vibratory or heat form. Thus by two transformations is heat transferred from one body to another by means of the ether, irrespective of distance.

Velocity of Ether Waves. — The ripples made by dropping a stone in water move with a velocity that depends upon the property of the water to transmit waves, not upon the manner in which the waves are made. Likewise the waves set up in the ether in any manner travel through it with a velocity that depends upon the properties of the ether. That they travel much faster than sound waves travel is plain, because one can see the light from a cannon or from fireworks some time before the sound of the explosion is heard. How fast ether waves or light does travel has been measured in a number of ways. One of them is by means of the motions of the satellites of the planet Jupiter. The satellites move about the planet with great regularity, and the time it takes for one revolu-

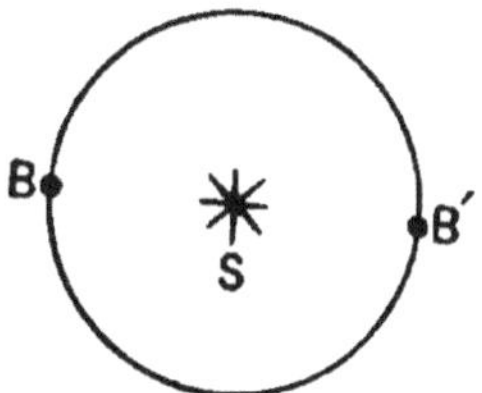

FIG. 94.

tion is very accurately known. In the figure 94 let **S** represent the sun, **B** and **B′** the earth in opposite parts of its orbit, and **J** the planet Jupiter, and one of its satellites in its orbit. When the earth is at **B′**, it is farther away from Jupiter than when at **B** by the diameter of the orbit of the earth, which is twice the distance of the sun from the earth, — 93,000000 × 2 = 186,000000 miles. The eclipse of the satellite is seen to take place 1000 seconds later at **B′** than at **B**, which shows that it takes the light that length of time to go from **B** to **B′**; that is, it goes at the rate of 186,000 miles a second through the ether, and a little slower through air, water, glass, and other transparent bodies. This very great speed of light or radiant energy will help one to conceive the vastness of the visible universe. Above and around the earth in every direction the stars are to be seen at night. It is the business of the astronomer to measure the distances to these luminous points, and he tells us the nearest of the fixed stars is so far away that it requires three and a half years for its light to reach the earth. Others are so remote that thousands of years are required. The light we now see from them left them before the Christian era began, before Rome was founded, or the pyramids were built, and if such a distant star were to be annihilated to-day, it would continue to shine upon us for thousands of years.

Wave-Lengths. — It has been found possible to measure the length of the waves that affect our eyes. Different colors have different wave-lengths, as in the following table:

Red,	the thirty-seven	thousandth	of	an	inch.
Orange,	" forty-two	"	"	"	"
Yellow,	" forty-five	"	"	"	"
Green,	" forty-eight	"	"	"	"
Blue,	" fifty-three	"	"	"	"
Violet,	" sixty-three	"	"	"	"
Lavender,	" sixty-five	"	"	"	"

There are waves longer and shorter than these, but the eye is affected only by those within the above limits.

Number of Vibrations. — One may compute the number of times the molecule vibrates per second when the velocity of the waves is known, and their length;

for, let v represent the velocity, l wave-length, n number of vibrations: $n = \frac{v}{l}$. If l be the wave-length for red waves $= \frac{1}{37000}$ of an inch, then n will be the number of times $\frac{1}{37000}$ of an inch is contained in 186,000 miles, which is about 400 millions of millions of times, and for waves $\frac{1}{65000}$ of an inch long, about 750 millions of millions. A pocket tuning-fork may vibrate 522 times a second; the shortest piano-string, 4000 times; the highest sound that can be heard is made by bodies vibrating 25,000 times; but these are hardly to be compared with the amazing number of vibrations made by atoms and molecules when they produce ether waves. The smaller a body is, the higher is its rate of vibration. A tuning-fork the fifty-millionth of an inch in length, if made of steel, would vibrate about 30,000 millions of times a second; if it were made of ether instead of steel, its rate of vibration would be greater in the proportion that the velocity of movement in ether is greater than that in steel, which is about 50,000 times.

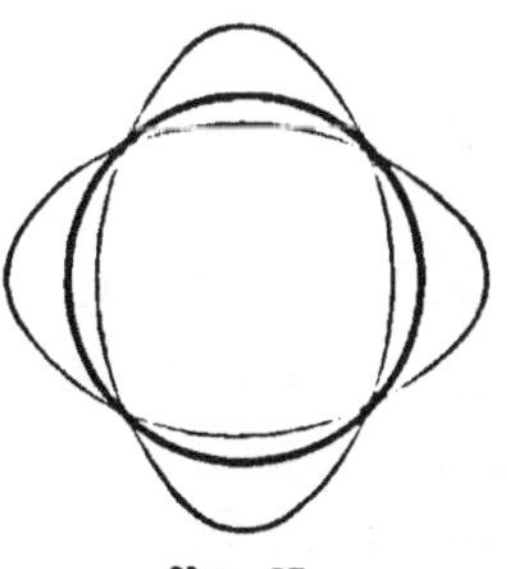
FIG. 95.

50,000 × 30000,000000 = 1,500,000,000,000,000, a number which shows that the vibratory motions of molecules, as determined by their wave-lengths, are such as would be expected of such minute bodies if they were made of ether.

The ether waves we call light are to be thought of as traceable back in every case to molecules and atoms that are in a true state of vibration; not oscillations to and fro, but changes of form such as a ring of brass or steel wire 5 or 6 inches in diameter would make if plucked (Fig. 95).

Phenomena of Ether Waves. — A line of waves like **a b** (Fig. 93) is called a *ray*.[1] A bundle of rays is called a *beam*. The rays from a candle or other source of such waves may be seen from every direction in an apartment, and every part of the flame may be seen, so every part of the flame must be giving out rays in every direction.

A beam may consist of *parallel*, *converging*, or *diverging* rays. Hold a convex lens in the sunshine, and the

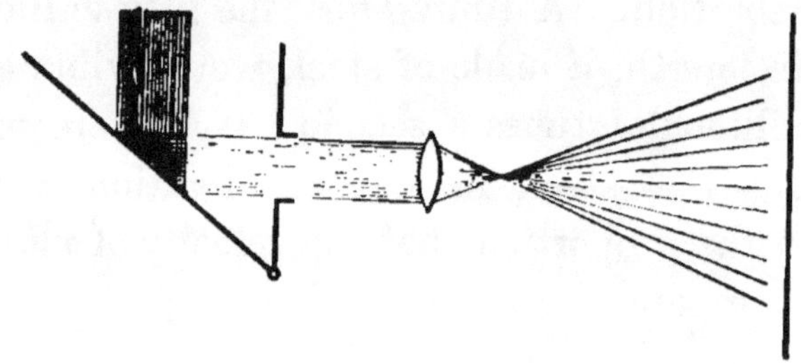

FIG. 96.

rays will be converged to a point from which they will diverge. If this be tried in a darkened room with a beam of sunshine directed into the room by a mirror outside (Fig. 96), the direction of the beam may easily

[1] The definition of a ray as here given is only a convenient term to indicate the direction of the forward movement of ether waves. In reality there are no more such individual things as light rays than there are water rays when a pebble is thrown into water, and waves travel out in all directions on the surface. A line drawn radially from the center of disturbance would go through all the waves, and the waves would travel through the line prolonged. Such a line might be called a ray.

be seen by means of the dust particles in the air. A little chalk-dust serves well to show the character of the beam. Before the beam enters the lens, it is parallel; after passing through, it is converged to a point beyond which it diverges. The point of light in front of the lens may be used as a luminous point, and objects held between it and the wall will have sharply defined shadows.

The shape of the shadows shows that the *rays go in straight lines;* otherwise the shadows would be irregular. If the source of light has considerable area, the edge of the shadow cast by it will not be as dark as the middle of the shadow. Hold an object like a ball in the light of a candle or lamp, as in Fig. 97, and observe the

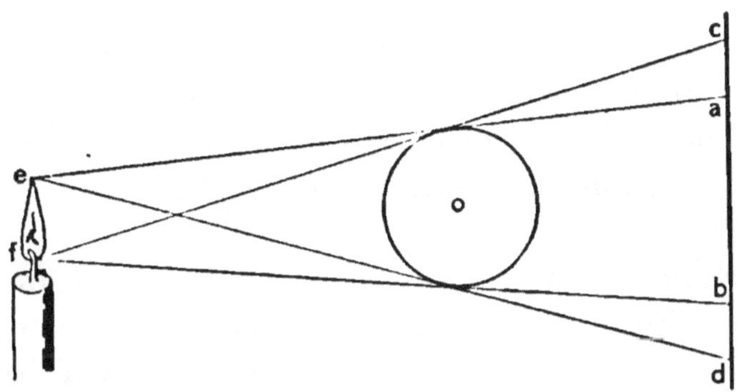

FIG. 97.

character of the shadow. The light from the tip of the flame casts its shadow from **a** to **d**; the light from the base **f** gives its shadow from **c** to **b**. Between **a** and **c**, **b** and **d** some rays fall, but none between **a** and **b**. The part **a b** is called the *umbra*, or complete shadow; the other is called the *penumbra*, or partial shadow. The shadows of objects in the sunshine are never well defined; the luminous surface of the sun is so large there is always a penumbra.

If the shutters or the curtains of a room be drawn so as to exclude most of the light, and an aperture half an inch in diameter be made in one of the shutters, a picture of the whole landscape will be visible upon the walls and ceiling of the room. By holding a sheet of white paper 2 or 3 feet away from the aperture, most of the prominent outside objects will be recognizable, everything in its proper place and color, but all inverted. This effect is finest in winter when snow covers the ground, as so much more light is reflected from snow than from grass or common objects. The image is inverted because the light rays all go in straight lines, as indicated in Fig. 98.

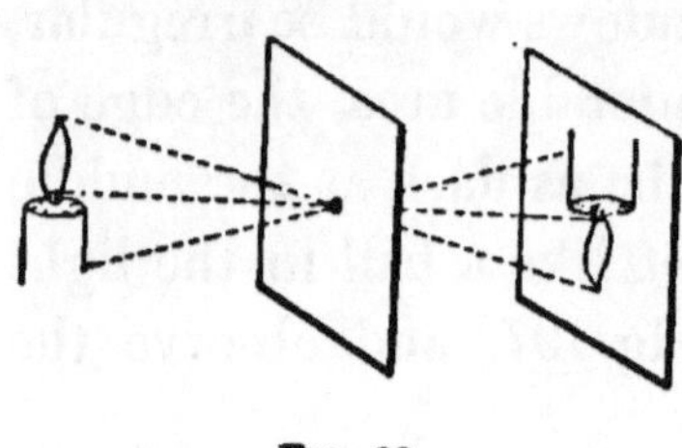

FIG. 98.

Brightness of Illumination.—If a certain amount of light from a point falls upon an area of a square foot at any distance from the point, the surface will have a certain degree of brightness. If it be removed to twice its original distance, only one-fourth as much light will now fall upon it, for the same amount of light will be spread over four times the surface. This follows because light moves in straight lines (Fig. 99). The brightness will therefore be but one-fourth the brightness of the surface in the first position.

This is sometimes expressed by saying that the intensity of light varies inversely as the square of the distance from the source. The relative brightness of two sources of light may be measured by observing the difference in the density of the shadows produced by each (Fig. 100). The shadows of the post upon the screen surface may be made to fall side by side by moving one or the other sources of light. By moving either of them towards or away from the post, a place may be found where the two shadows are apparently alike. If the distance from the screen to each source be now measured, the relative brightness of the sources will be as the square of their distances. Suppose a candle be placed 1 foot away from the screen, and a lamp is found to give an equal shadow at the distance of 4 feet; then the latter will be as many times brighter than the other as the square of 4 is greater than the square of 1; that is, the lamp gives a light equal to 16 candles. Such a device is called a *photometer*. The standard of brightness is commonly that of a sperm candle burning 120 grains an hour. The common

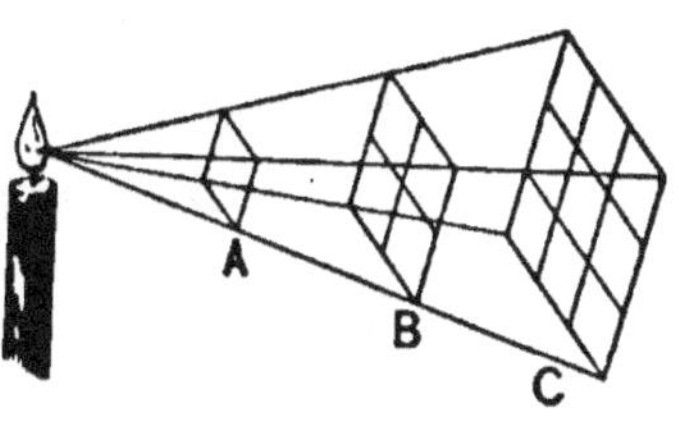

FIG. 99.

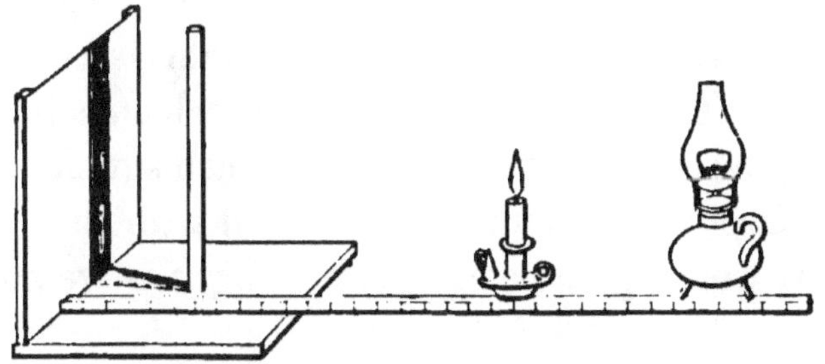
FIG. 100.

paraffine candle differs but little from this. A lamp that gives 10 times as much light as this standard candle is said to be of 10 candle-power.

The following table gives the relative brightness of several common sources of light:

Standard candle,	1	candle-power.
Kerosene, Rochester lamp,	25	" "
Gas jet,	12 to 18	" "
Gas, Welsbach burner,	30 to 50	" "
Oxyhydrogen lime light,	200	" "
Electric arc, common,	800	" "
Electric arc search-light,	1,000,000 or more	" "

The sun does not much differ in brightness from the electric arc, which is the brightest that can be produced artificially.

Brightness and amount of light are different things. A spark may be brighter than a candle flame, but the latter gives more light because its shining area is greater. The moon gives much light because it is so large. The part of an electric arc which is brightest is rarely more than the tenth of an inch square; but if the surface of the sun were covered with such bright spots, it would give nearly as much light as we now get from it.

In these and all other sources of light depending upon combustion, there are chemical actions taking place, and atoms are recombining into new molecules. Whenever this happens, the atoms are vigorously shaken and vibrate with great energy. This energy they lose by radiation. When a current of electricity produces light in a carbon filament, the carbon molecules are highly heated and therefore vibrate; they, too, lose their energy by producing ether waves, the energy of the waves being equal to the energy supplied, else the temperature of the lamp would not remain uniform.

Action of Matter upon Ether Waves.—When ether waves meet a mass of matter they are either *reflected*, *absorbed*, or *transmitted;* usually all three happen in some degree, depending upon the kind of matter and its condition.

When the direction of the rays is changed, as when a ray of sunlight **a b** (Fig. 101) falls upon a mirror, the rays **b c** are said to be *reflected*. A ray or a beam which falls upon any surface whatever is called an *incident* ray or beam. Thus **a b** (Fig. 101) is an incident ray; **b c** is a reflected ray.

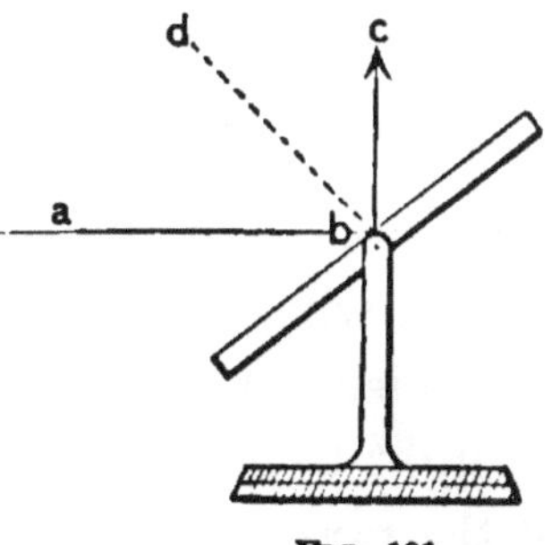

FIG. 101.

If a line **b d** be drawn perpendicular to the surface of the mirror at the point **b** where the ray meets it, then the angles **a b d** and **d b c** are equal; **a b d** is called the *angle of incidence*, and **d b c** is called the *angle of reflection.* These angles are both in the same plane. Let the mirror be tilted so the reflected ray will be directed back towards **a**, then **d** will also be in the same line. If the mirror be tilted forty-five degrees, the ray of light will have moved ninety degrees, and in general the ray will move through twice the angle the mirror moves through.

If in a darkened room a small beam of sunlight be

admitted from a mirror outside, and a mirror be held in its path, it will have its direction changed, and may be traced through the air to the wall. If a piece of white paper or other light-colored object be put in the path of the beam, it will reflect light to every part of the room. The difference in action of the glass and the paper is due to the difference in the character of the surfaces of the two bodies, the glass being smooth and even, the other very uneven (Fig. 102). A common magnifier will show that the surface of the paper is rough, so that the individual rays meet it at different angles, where each follows the above law of reflection.

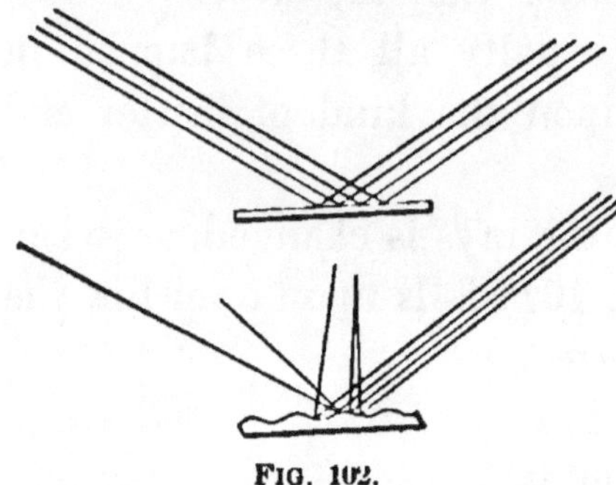

FIG. 102.

A piece of clear glass or mica will allow such a beam of light to go through it without much loss; but some of the rays will be reflected to the wall, and the amount of these so reflected depends upon the angle the glass presents to the beam. If the beam be perpendicular to the surface, the least amount will be thus reflected. If the beam meets the surface at a large incident angle, nearly all the light will be reflected and but little transmitted. A body that permits light to go through it so that objects may be plainly seen through it like common window glass is called a *transparent* body. One that scatters the light that goes through it like paper or roughened glass is called a *translucent* body; and one that prevents all light from going through is called an *opaque* body. If the opaque body have a dark and roughened surface, like a blackboard, but little light will be reflected from it. The light will be absorbed; as radiant energy it will be changed back into heat, and the temperature of the opaque body will rise. The warmth felt upon the face and hands when exposed to sunshine or the light from a fire shows this, and many objects become uncomfortably hot when left in the sunshine for a time, because they have absorbed the energy of the ether waves that have fallen upon them. A perfectly transparent body would not get thus heated, for if the ether waves were transmitted, their energy would not be left behind.

A piece of clear glass or a quantity of clear water becomes warm when exposed to sunshine or firelight, which shows that they are not transparent for all waves. The clearest lamp chimneys stop about ten per cent of the light which falls upon them; and at the depth of one hundred feet in clear water it is dark, though the sun may be shining vertically into the water. All the light and all the energy, except what is reflected at the surface, is absorbed. If there were a body which would absorb all waves, it could not be seen unless it were self-luminous. The blackboard absorbs more light than chalk absorbs.

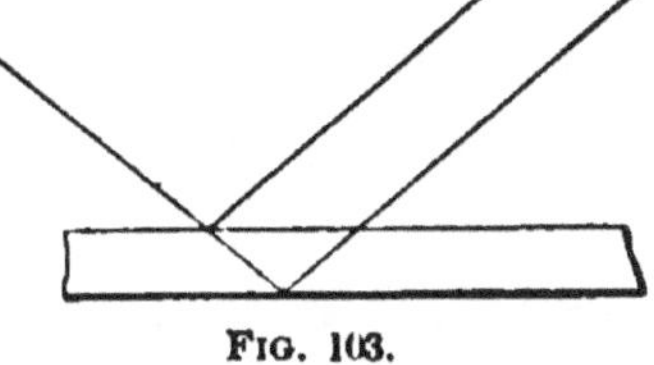

Fig. 103.

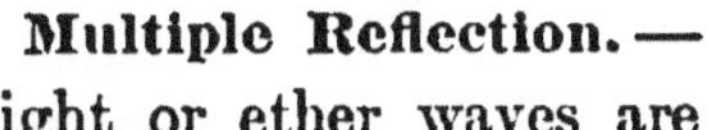

Multiple Reflection. — Light or ether waves are reflected in some degree from every surface they meet. A piece of plane glass held in the sunlight reflects from its front surface much of the light falling upon it. If the back of the glass is covered with mercury or silver, we have what we call a looking-glass, which reflects nearly all the light, but the light is reflected from the second surface. If a piece of thick plate glass be held so a small beam of light half an inch in diameter can fall obliquely upon it, a number of reflections can be seen on the wall, caused by the reflections back and forth through the glass at the surface (Fig. 104). These are called *multiple reflections*. The thicker the glass, the farther apart these will be; and the greater the angle of inclination of the beam, the brighter the reflections will appear, as a larger amount of light is reflected at each surface. In looking into a plate-glass mirror one may sometimes see a faint double image of himself, the fainter one being the reflection from the front surface of the glass.

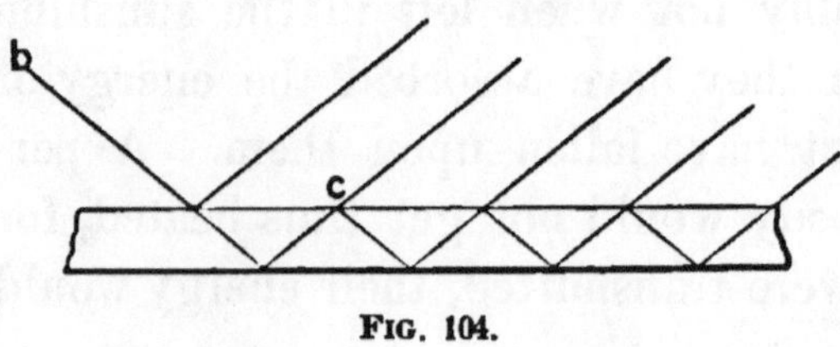

FIG. 104.

MIRRORS.

I. Plane. — An object seen by reflection from a mirror appears as far behind the mirror as the object is in front of it. If the object be seen by reflected light, then the light which comes to the eye must have been reflected from the surface of the mirror at such points as lie between the eye and the image as seen. The position of the image and the direction in which it is seen may be drawn correctly by first drawing the parts of the image as far behind the mirror **M** (Fig. 105) as they are front of it, as at i, then drawing lines from the extremities of the image i to the eye **e**. These two lines represent the direction of the rays of light after reflection from the mirror. If lines now be drawn from the extremities of **o** to these points on the mirror, they will represent the direction of the rays upon the mirror which are reflected to the eye at **e**. Other eyes at **a** or **b** would still see the image **i** in the same position, but different rays from **o** would be reflected from different points on **M**, and would be drawn in same way without change of position of the image.

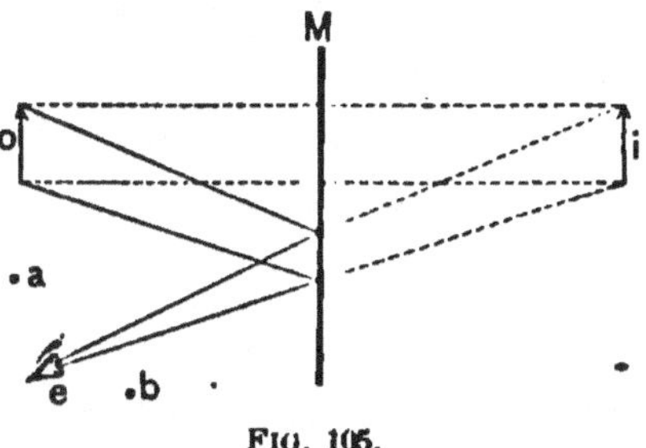

FIG. 105.

II. Curved Mirrors.—Reflecting surfaces may also be made concave or convex; but whether the one or the other, the law of reflection holds true for them. There may be any number of kinds of curved surfaces. Let **a b** (Fig. 106) represent a concave mirror,—a part of a spherical surface,—with **c** as the center of curvature. Suppose that at **c** there were a source of light so that radiations would travel in every direction from it. Any rays, **ce**, **ch**, or **cl**, would meet the surface of **ab** at right angles to it, for these lines are radii; these rays and

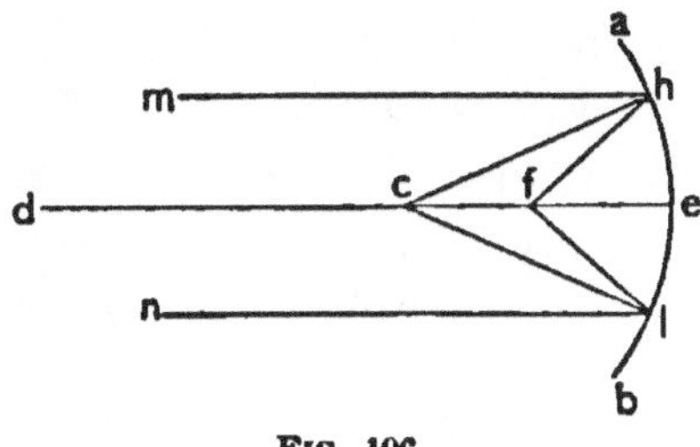

FIG. 106.

all others would be reflected back to **c**. Suppose a ray from some other source and direction as **m h** strike the mirror at **h**. It will be reflected so that the angle of incidence equals the angle of reflection; **m h c** is the angle of incidence, for **c h** is at right angles to **a b**. Draw **c h f** = **m h c**; **h f** represents the direction of the reflected ray. It crosses the line **d e** at **f**. In like manner the ray **n l** is reflected to the same point on line **d e**. All parallel rays will be reflected to this same point **f**, which is half-way distant from the mirror to its center of curvature. This point is called the *principal focus* of the mirror; that is, it is the focus for parallel rays. If now a source of light were to be placed at **f**, the light would be reflected in a beam parallel to **d e**, and would travel on to an indefinite distance.

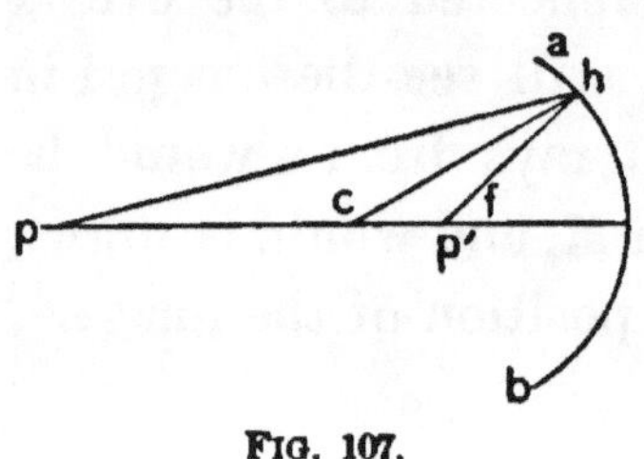

FIG. 107.

The axis of this beam **d e** is called the *principal axis* of the mirror. Let another ray start from some other point on this axis as at **p** (Fig. 106), and fall on the mirror at some point, say at **h**. The angle **p h c** would be less than **m h c** of Fig. 107; hence the reflected ray would cross **c e** at **p′**, between **f** and **c**, and the closer **p** is to **c**, the nearer will the reflected ray be to **c**, until points **p** and **p′** coincide with **c**. A source of light at **p′** will be reflected to **p**; hence **p** and **p′** are called *conjugate foci*.

Images Produced by Concave Mirrors. — If a candle flame be brought up near the principal focus of a concave mirror, a place may be found by trial where the image of the flame may be seen upon the opposite wall. It will be inverted and enlarged. If without moving the mirror the flame be carried to the wall where its image appeared, and a small piece of white paper be put where the candle was, a small inverted image of the candle will appear upon the paper. These points represent the conjugate foci mentioned above. A diagram showing how these effects can be made may be drawn as follows: Let **a b** (Fig. 108) represent an object

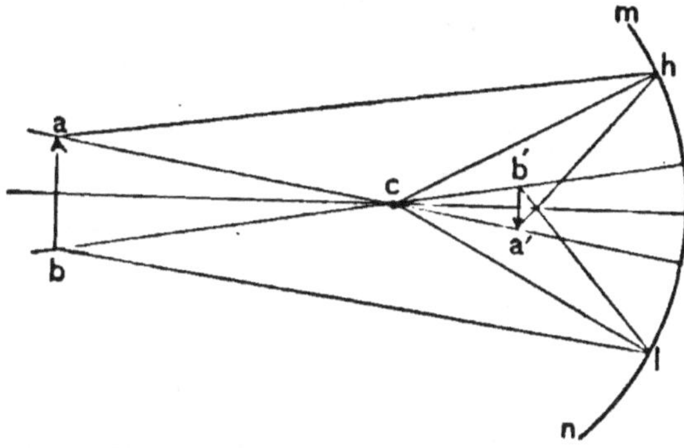

FIG. 108.

at any distance from the mirror **m n** beyond the center **c**. Light from every portion of the object falls on every part of the mirror. A ray **a c** through the center of curvature of the mirror will be reflected back in the same line; likewise the ray **b c**. Draw a line from **a** to any part of the mirror as to **h**. The reflected ray will cross the first ray, which goes through the center of curvature at **a'**, and all rays from **a** will be reflected to this same place below the axis. This will be the focus for the point of the arrow. From **b** draw a line to any part of the mirror as to **l**, making the angle of incidence equal to the angle of reflection; the reflected ray will cross the line through **b** and **c** at **b′**, and consequently the heel of the reflected arrow will be above the axis, and the image of the intermediate parts of **a b** will be between these two fixed points. The image will be inverted and smaller than the object.

An object **a b** placed between the center of curvature and the mirror has an enlarged image **a′ b′** apparently behind the mirror. The position and proportional size of this image may be determined by drawing lines from the center of curvature **C** past the extremities of the object **a b**, and prolonging them behind the mirror. Lines drawn from the same points of **a b**, and made parallel with the axis, will meet the mirror at **c** and **d**, and other lines drawn from the focus **f**, half-way between the center of curvature and the mirror, through the points **c** and **d**, and prolonged until they meet the first lines, will give the position of the image **a′ b′**. It is to be kept in mind that the parallel rays **a c** and **b d** will be reflected to **f**, as if they had come from **a′ b′**. Objects looked at in a convex mirror appear diminished in size, for **a b** and **a′ b′** are conjugate; that is, either may be the image of the other.

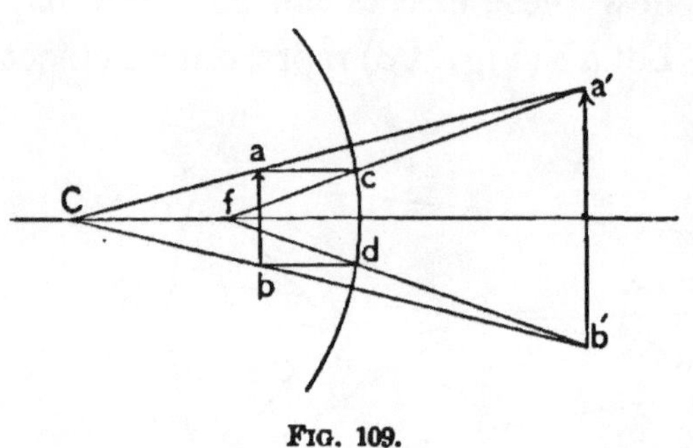

Fig. 109.

Refraction. — When a small beam of light is directed into water or a piece of glass at any angle except a right angle, the beam is observed to be bent from its original course, becoming more nearly perpendicular. Thus the beam a on reaching the glass is partly reflected from the front surface, as indicated by the arrow, and partly transmitted through the glass in a new direction. On reaching the opposite surface it is again bent into its original direction, provided the sides of the glass are parallel. This change in the direction of light on entering a new medium is called *refraction*, and the power of refraction is possessed in some degree by all transparent bodies, whether solids, liquids, or gases.

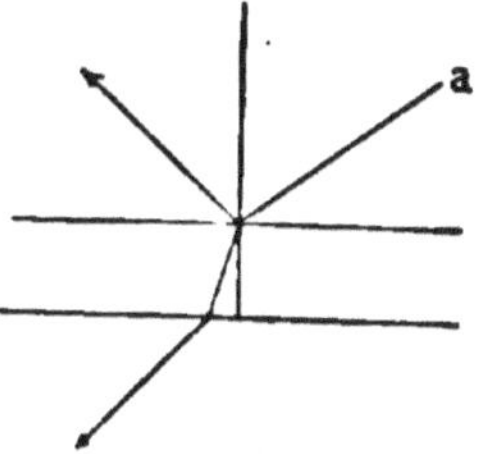

FIG. 110.

The new direction given to any ray is determined at the *surface* of the new medium, and does not change so long as the substance remains uniform in density and constitution, the ray continues to move in a straight line through it. If the substance does change in either density or composition, the ray is at once refracted more or less. Thus the light coming from the sun or a star has to traverse the thickness of the atmosphere, which is denser the nearer it is to the surface of the earth. The refractive power becomes greater and greater, so the real course of all rays except those coming vertically down, is a curved line, being more and more deflected. In the atmosphere this curvature is at a maximum when the source of light is at the horizon, and is sufficient then to tilt the ray downward half a degree, which is equal to the diameter of the sun or moon. It follows that either of these luminaries may be seen when they are really below the horizon.

The Law of Refraction.—For a given substance like water or a piece of glass, it is found that the change in the direction of a ray of light may be known when the angle **a c b** (Fig. 111) at which it meets the surface is known; **a c b** is the angle of incidence, **e c d** is the angle of refraction. Describe a circle with **c** as a center, and from points on the circumference where **a c** and **c e** cross it, draw lines perpendicular to **b d**, that is, **i f** and **e h**.

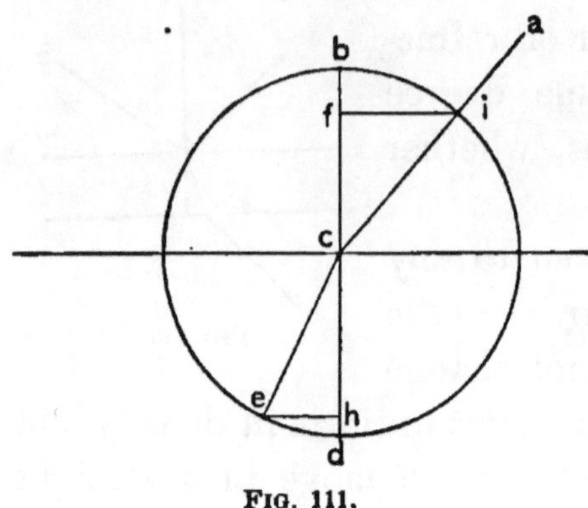

FIG. 111.

These lines represent the sines of the angle of incidence and of refraction, and $\frac{\text{sine } \mathbf{acb}}{\text{sine } \mathbf{ecd}}$ is the same whatever the value of **a c b**; it is, therefore, a constant quantity and called the *index of refraction.* If the light goes from air into water, the value of this index is $\frac{4}{3}$; if from air to glass, $\frac{3}{2}$.

There is another meaning to these figures: they indicate the relative rates at which light travels in the substances. Thus, if light travels in air at the rate of 185,000 miles a second (a thousand miles a second less than in free space), it travels but three-fourths as fast in water and two-thirds as fast in glass. This has been proved by experiment.

How Light becomes Refracted.—As light is a wave motion, it may be mechanically represented by shaking rapidly the end of a long, limp rope; the wave would travel along it, but the movement of each point on the rope is a to-and-fro motion at right angles to the direction of the wave, the same as the movement of the hand that starts the wave. In a free medium the direction of such a wave is *always at right angles to the direction of the vibration;* it is the latter that determines the direction of forward movement. Suppose **a b** represents the direction of march of a platoon of soldiers **d e**, keeping their front at right angles to their direction of march. So long as there is no obstruction at any part of the way, they will all move with uniform velocity. If they should come to more difficult ground **h d**, at an angle where the rate of march would be retarded, **d** would meet it first. When the middle of the platoon **b** had reached the same ground, **d** would not have been able to travel as far, but would have gone to **g**; when **h** reached the same ground, **d** would have got only to **f**, and **d f** is shorter than **e h**. The frontage of the whole body would have been changed to a new direction, and if the direction of march was to be perpendicular to the platoon, it would now go towards **c**; there would have been refraction. In similar way, if the line of march were reversed from **c** towards **b**, **h**

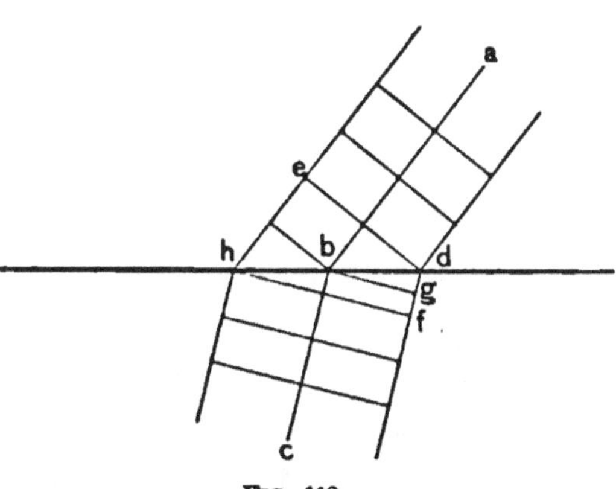

FIG. 112.

would get out soonest, and the whole body would be swung round to face a, and the rate of advance would be quickened. For such reasons the direction a ray of light takes on entering a new medium depends upon the angle at which it meets the new medium and upon density.

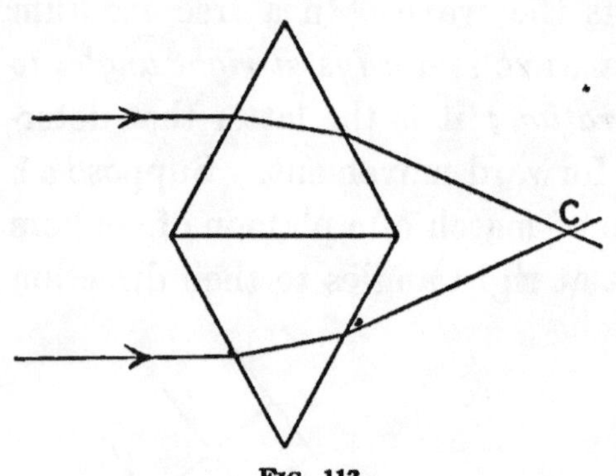

Fig. 113.

If a ray be made incident upon a triangular piece of glass called a *prism* (Fig. 113), it will be bent still more from its original direction on leaving the prism, as indicated by the arrows. Another prism inverted will direct another ray across the first at C. If instead of flat-faced prisms the glass be made with curved surfaces (Fig. 114), parallel rays will all be brought to the same point f, which is called

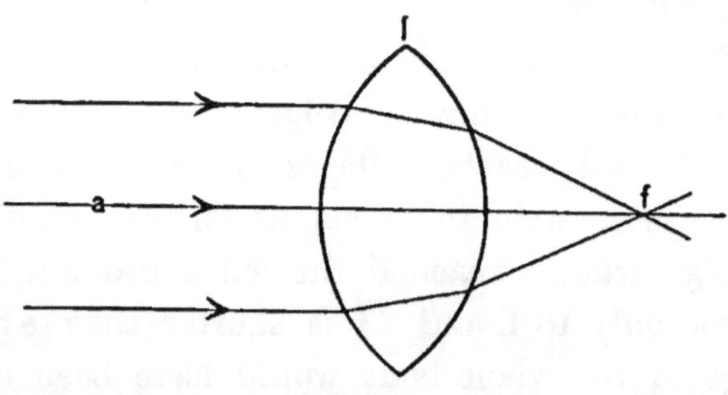

Fig. 114.

a *focus*, and the glass with such a curved surface is called a *lens*. Of course there may be any degree of curvature, and it may be convex or concave. A lens with only one side curved is called a plano-convex or plano-concave lens, as the curved surface is convex or concave. If both sides are curved, it may be double convex as l in Fig. 114, or double concave, or one face convex, the other concave, and is then called a *meniscus*.

Properties of Lenses. — A line drawn through the center of a lens as **P P′** is called its axis, and the focus of a lens is somewhere along this axis. Rays parallel to this axis as **a** and **b** come to a focus at some point on this axis as f, which is called the *principal focus* of the lens. *The focal length of the lens is the distance*

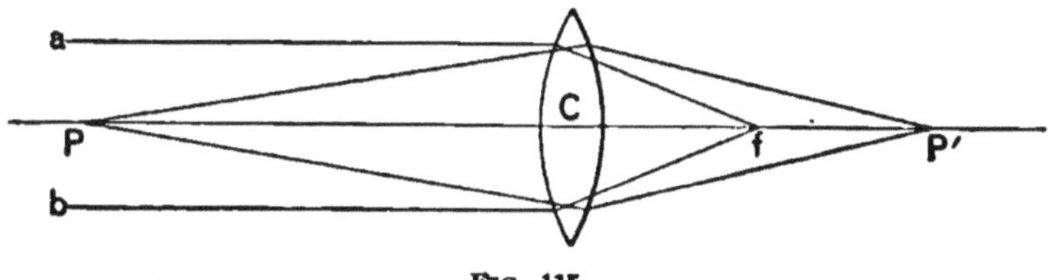

FIG. 115.

from the middle of the lens **C** *to the principal focus.* The focal length of the lens may be measured by holding the lens in the sunshine so as to focus the rays upon a piece of paper, and then measuring its distance from the lens; or in a darkened room hold the lens so an image of a distant object, like a tree or a steeple a mile away, may be plainly seen on a piece of white paper, then measure the distance from paper to lens. Nearer objects will give a different and larger value. Thus if **P** be a source of light, rays from it falling on the lens will be brought to a focus at **P′**, further away than the principal focus. Also, if a source of light be placed at **P′**, the rays will come to a focus at **P**. These two points so exchangeable with each other are called *conjugate foci.* Any change in the position of one produces a corresponding change in the other. There is a definite relation between the principal focus and the conjugate foci of a lens, represented by the expression $\frac{1}{p}+\frac{1}{p'}=\frac{1}{f}$, where **P** and **P′** are the conjugate foci and **f** the principal focus; so if one knows two of these factors, he may compute the other.

Images Formed by Lens. — Place a lens between a lighted candle or lamp and the wall where an image may be formed (Fig. 116). By moving either the lens or the

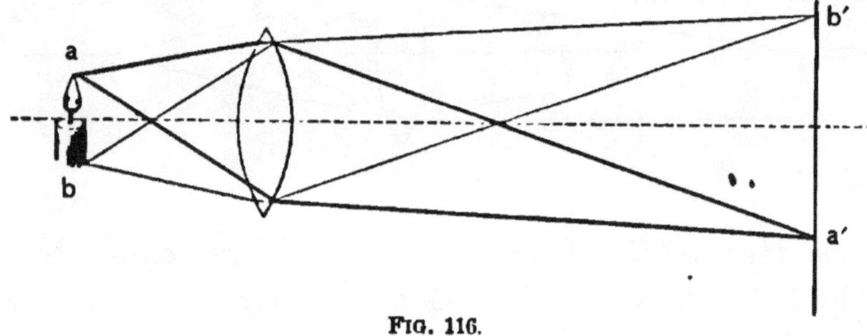

FIG. 116.

light along the axis of the lens, a point will be found by trial where an enlarged and inverted image of the flame will be seen upon the wall. This may be understood by remembering that for every point on **a b** there is a conjugate focus at **a′ b′**. The rays from point **a**, which fall on the lens, will be brought to a focus at **a′**. The lines in the figure are made heavier in order to make this plain. In like manner all the rays from **b** will be focused at **b′**. All points between **a** and **b** will have corresponding points between **a′** and **b′**. But **a′** is below the axis while **a** is above it. The distance of **a′** from the axis is also greater than the distance of **a** from it; that is, the image is larger — is magnified. If the distance from **ab** to the lens be known, as well as the focal length of the lens, the distance to the screen at **a′ b′** can be computed. Thus suppose focal length of lens be one foot, the distance of **ab** from lens 14 inches; how far away is the image?

$$\frac{1}{p}+\frac{1}{p'}=\frac{1}{f} \quad \frac{1}{14}+\frac{1}{x}=\frac{1}{12} \quad x=7 \text{ feet}=\text{distance}$$

of the inverted image.

This use of the lens is very common, and is called *projecting.* Either sunlight, lime light, or electric light

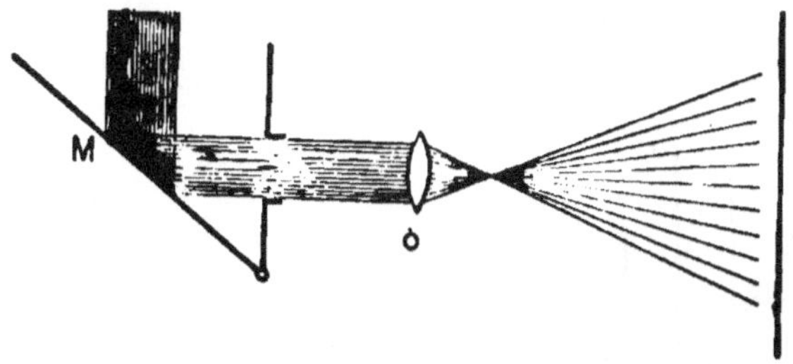

FIG. 117.

may be used. For sunlight a mirror **M** is fixed in the window so as to direct a beam of sunlight into the room onto a screen. This is called a *porte-lumière.* A double convex lens **o**, four or five inches in diameter and about a foot focus, answers well. This placed in the path of the beam serves to give a disk of light on the screen. The size of the disk depends only upon its distance away from the lens. Any object placed about a foot back from the lens will show an enlarged image on the screen, and transparencies made for such purpose placed there will give bright and pleasing pictures, better than can be produced by the electric light. A lens used for this purpose is called an *objective.* For still better definition than such a single lens can give, objectives are sometimes made compound.

By drawing the lines to indicate the direction of rays from a lens, it will be seen that the shorter the focus of the lens, the more the light will be dispersed, and consequently the less bright will a picture be. To make it brighter, more light must be used. This is effected by using a lens as a light condenser. An object at **a** (Fig. 118) has more light upon it than it would have had nearer the lens **c**, which when thus used is called the

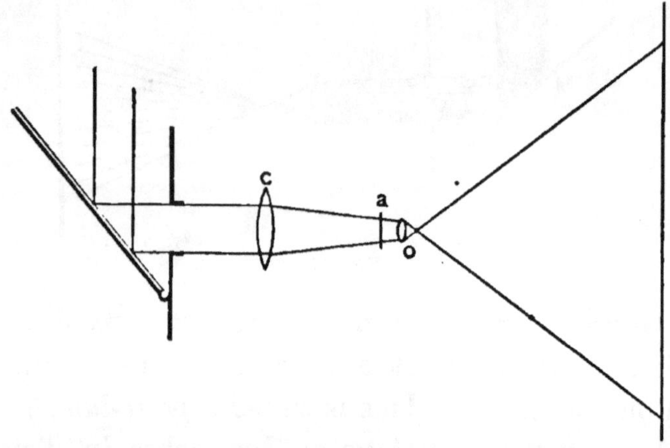

FIG. 118.

condenser. The short focus objective **o** magnifies the object very much and enables one to see microscopic things, such as vinegar eels, by placing them in a glass tank and holding it at **a**. A plate of glass placed there, having a small drop of solution of sal ammoniac rubbed on it, will finely show the process of crystallization. This device is called the *solar microscope.*

The Lens as a Magnifier. — If one looks through a convex lens at an object nearer to the lens than its focus is, the object appears enlarged. Let **a b** be an

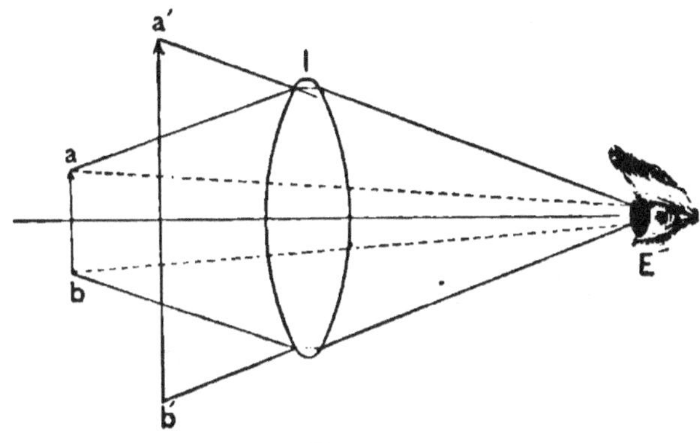

FIG. 119.

object. To an eye at **E** its magnitude would be measured by the angle **a E b**, which is called the *visual angle.* If a lens **l** be placed between the object and the eye, the ray **a l** will be refracted to the eye at **E**, and will make the point of the arrow seem to be in the direction of **E l**; the same for the rays from the heel of the arrow. This makes the visual angle **a′ E b′** much wider and the object looked at appear much bigger and nearer; that is, it has been *magnified.* Such a lens is called a *simple microscope.*

The Compound Microscope.—To see minute objects a small, short focus lens is needed. As it cannot be brought near enough to the eye to conveniently see with it, it is used in combination with another and larger lens. The smaller lens **o** next to the object **a b** forms an image of the latter at some point as at **c d.**

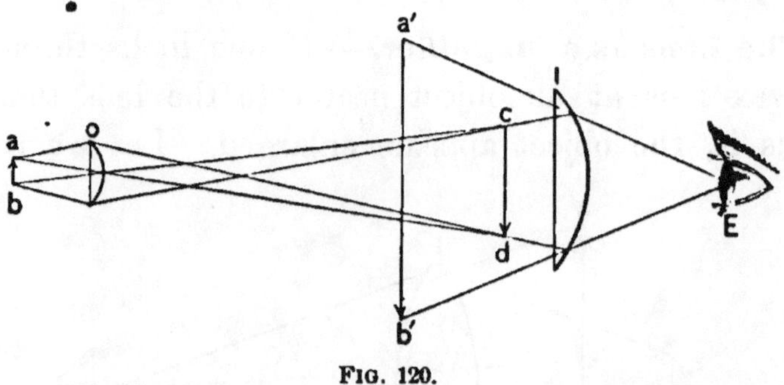

FIG. 120.

The same rays pass on to the larger lens **l**, called the *eye piece*, and by this they are refracted to the eye, making a much wider visual angle than there could have been otherwise, and **a b** is seen amplified to **a' b'**.

FIG. 121.
Compound Microscope.

For convenience these lenses are mounted in tubes capable of adjustment (Fig. 121). The objective **o** of a microscope is generally compound, and its focal length is generally less than an inch, and may be as short as one-tenth or one-twentieth, but satisfactory work cannot be done with such short ones. A good microscope may enable one to see an object as small as a hundred-thousandth of an inch in diameter. The magnifying power of a microscope is generally stated in diameters, that is, as 100 or 500 diameters. An object $\frac{1}{100}$

of an inch long in actual measure would appear to be one inch long if magnified 100 diameters.

If the diameter of a molecule of water be the fifty-millionth of an inch, and the best microscope of to-day will show nothing finer than the hundred-thousandth of an inch, then in order to see the molecule it is needful to make the microscope as many times better than we now have it as 100,000 is contained in 50,000,000, which is 500 times. Then the molecule would appear simply as a point without parts. But motions of a body are magnified as much as their parts, and if molecules be in incessant motion, as their phenomena indicate, they could not be seen if the microscope were otherwise able to show it. Hence there is no probability of one's ever being able to see such a molecule.

The Telescope. — The telescope is a combination of two or more lenses to enable one to see distant objects more plainly, *viz.*, a large lens **o** (Fig. 122) for forming

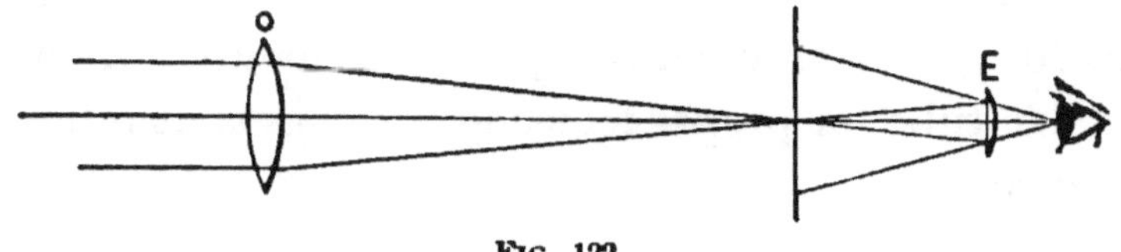

FIG. 122.

an image of a distant object at its focus, and a smaller lens for the eye piece **E**, to receive and again refract these rays to the eye ; the image is enlarged to a degree that depends upon the magnifying power of the eye piece **E**. The image is inverted as already explained. For looking at the stars this makes no difference, so for an astronomical telescope these two lenses only are necessary. The larger the objective, the more light reaches the eye, and the better a faint and distant object can be seen.

The greater the diameter of the objective, the longer is its focal length and the tube needed for its use. An objective 36 inches in diameter needs a tube 60 feet long, and the 40-inch objective for the Yerkes telescope at Chicago has a tube nearly 70 feet long. Only about 2000 stars can be seen on a clear night by the eye alone; but in the whole sky as many as a hundred millions can be seen with the large glasses, and each increase in the size of these glasses reveals other stars beyond in every direction.

The Spy Glass. — When two glasses are used the image seen is always inverted. This renders it unserviceable for common use. By employing two more glasses the image is again inverted, thus bringing it to a proper position; and so a spy glass is provided with four glasses, the first pair for magnifying, the second pair for giving an erect image, — all mounted in an adjustable tube for proper focusing for objects at different distances.

Prismatic Refraction. — When a beam of sunlight is sent through a triangular prism so as to be refracted, it is noticed that the beam leaves the prism, not colorless as it entered, but as a band of colored lights (Fig. 123), which show upon the wall as a series of rainbow tints in which as many as six distinct colors may be counted, — red, orange, yellow, green, blue, and violet. The red is least refracted while the violet is most refracted. If a similar prism be held in either of these colored rays, the rays will be still more refracted, but none of them will be broken up into other colors. These colors in this order are known as the *spectrum*. The same series of colors may be seen by letting light from any common source of light, as a candle, gas-jet, or an electric light, pass through a prism. If now a second prism be placed against the first prism so all the light will have to pass through it also (Fig. 124), the whole beam will be refracted back to the straight line it would take if no prisms were in its path, and the colors will disappear, leaving a white spot upon the wall opposite the window. These experiments show that white light from whatever source is compound, and the prism serves to separate the constituents, which may be again reunited into white light. In various ways the rays of these colors have had their wave-lengths measured, and some of the measurements are given on page 208. By comparing the positions of the tints in the spectrum with these numbers, it will be seen that *the shorter the waves, the more they are refracted.* This means that a prism separates the rays of light in the order of their wave-lengths.

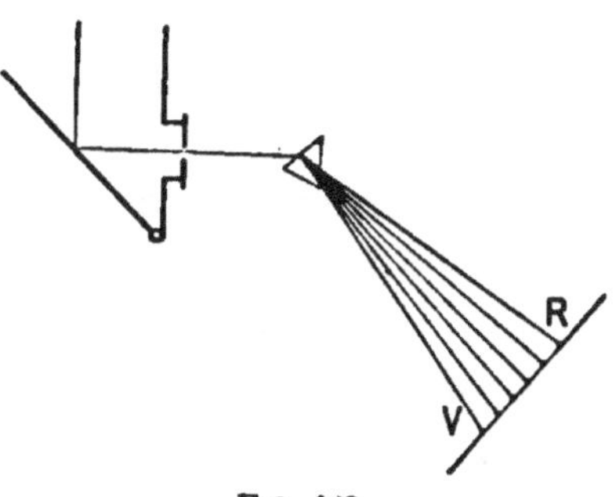

FIG. 123.

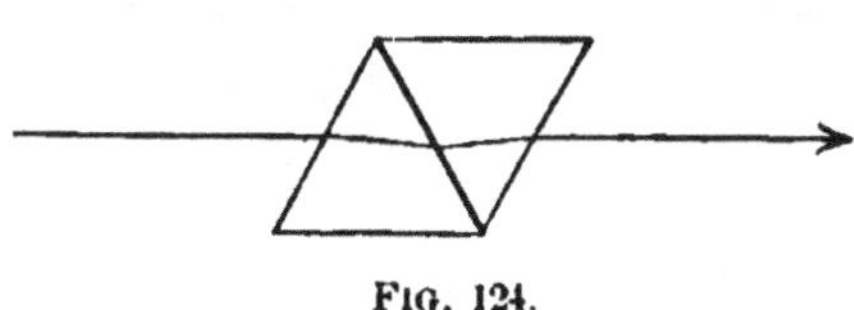
FIG. 124.

In order to see these colors to the best advantage, it is better to employ a slot through which the sunlight is directed (Fig. 125). It may be an inch long and the sixteenth of an inch broad. A slit cut in paper will answer if a metallic adjustable apparatus is not at hand. In the path of this narrow beam when it enters the room, place a lens of ten or twelve inches' focus and move it in the beam until a sharply defined and enlarged image of the slit is seen on the opposite wall; then place the

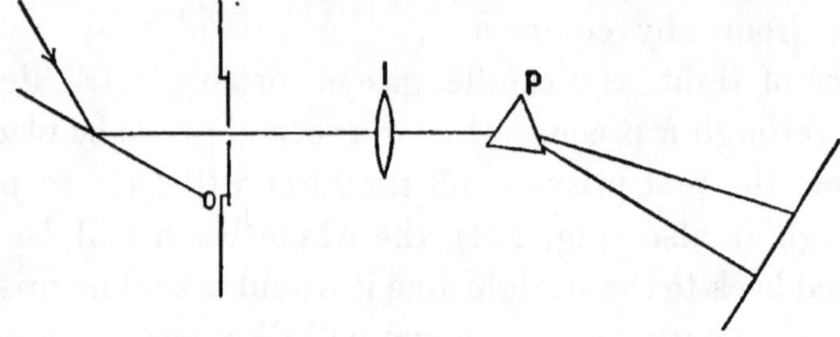

FIG. 125.

prism in front of the lens and near it. The beam will be bent to the side of the room and a spectrum will appear, the length of which will depend upon the distance to the wall or screen and also upon the kind of prism used. A dense glass prism should give at 20 feet a spectrum about 3 feet long. A bottle prism of bisulphide of carbon will give one 4 feet long. By turning the prism this way or that, a position will be found where the deflection of the whole is least, and if the prism be turned either way from this, the spectrum will move away in the same direction. This deflection is called the *angle of least deviation.* Prisms of crown glass, of quartz, of salt, or water, when tested in this way show that for each there is an angle of least deviation different from any of the others.

Spectrum Analysis. — The colored fires seen in fire-works are produced by igniting some of the salts of the metals. If a loop of platinum wire be dipped into salt water, and then be held in the flame of a Bunsen burner or an alcohol flame, the flame will be colored bright yellow. If the wire be dipped into a solution of lithium chloride, it will color the flame red; potassium chloride

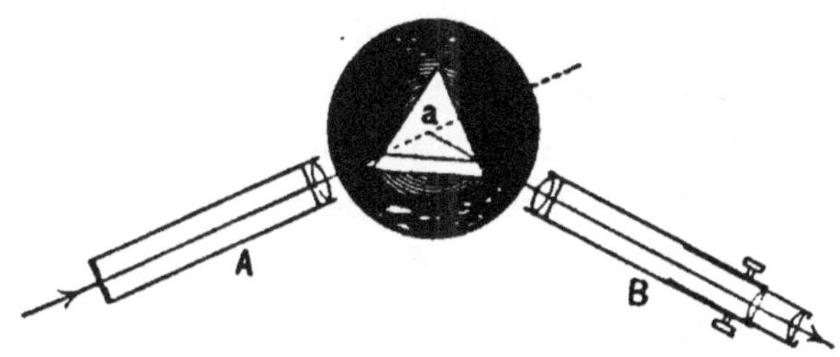

FIG. 126.

will color it purple; copper, a greenish blue. These different colors signify that different wave-lengths are produced by the vibrating molecules. If such light be sent through a prism, it is found to give, not a spectrum like the sunlight, but a spectrum of its own, peculiar to each element, and like no other. The better to observe these an instrument has been devised called a *spectroscope* (Fig. 126), consisting of a tube **A** carrying a slot, through which the light goes to the triangular prism of flint glass. This disperses the waves in the order of their wave-lengths, and they pass on into a short telescope **B** for conveniently examining the spectrum produced. Thus it is found that sodium gives one yellow image of the slot, and it is called a yellow line. Lithium gives a red line. Copper gives blue and green lines. The bright lines produced in this way are called the *spectra of the elements.* So it is possible to make a chart of the spectra of all the elements. The wave-length of the yellow light given out by sodium is about the forty-four thousandth of an inch; that of the red light of lithium, the forty-two thousandth of an inch. Most of the chemical elements give out rays of many different, but of definite, wave-lengths; but whatever number they give, the wave-length is uniform. The yellow light of sodium and the red light of lithium have always the same positions in the spectrum. This means that each kind of a molecule produces waves of a given length, and a mixture of different elements will not prevent them from doing so when in a flame. Each one will give its own spectrum independent of the rest.

If the light from an electric arc be examined in the same way as the sunlight, by sending it through

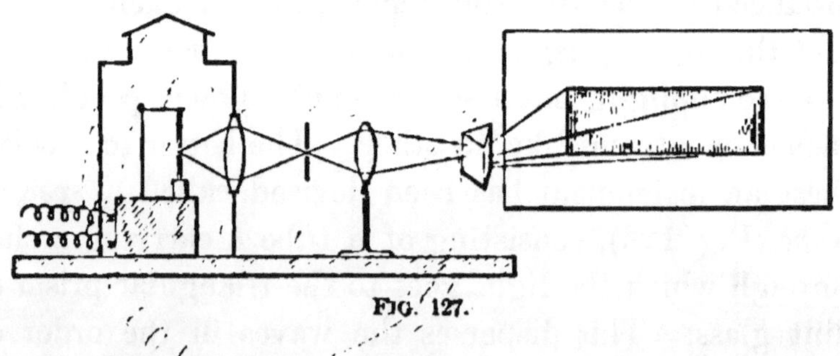

FIG. 127.

a slot in a lantern and then through a prism (Fig. 127), it is found to give a spectrum like the sun, all the colors in the same order. It gives a *complete* spectrum instead of a partial one, as is given by the substances just described. This is true for all kinds of solid bodies which are made hot enough to shine and give out light waves; also for glowing liquids like melted iron or other metals.

The explanation of this great difference between the spectra of solids and liquids and of gases requires the reconsideration of the molecular conditions in these different states of matter. Suppose a large number of bells were shaken together in a basket, there would be a great jingle of sounds, but no one of the bells could give out its own sound because its vibrations would be interfered with by its touching neighbors. In solids the molecules cohere even at a high temperature, and being practically in contact with each other, their vibratory motions continually interfere with each other. At a high temperature they are all vibrating at rates which give out light waves, but no one of them has time to complete a vibration before it is interfered with by the bumpings of the adjacent molecules which compel it to vibrate irregularly, in all periods, and thus produce irregular wave-lengths, some long and some short, which follow each other. When rays made up of such irregular waves come to a prism which separates the waves in the order of their lengths, they are spread out into the spectrum, and we see a *continuous* one, made up of the ether waves of all lengths which can affect the eye, and many more both longer and shorter that cannot affect it. Such solid bodies must produce waves of all lengths; whether they are composed of one element or another does not matter. The result is due to their compactness. In liquids, too, the same interference takes place among the vibrating molecules when they are heated to incandescence. They therefore give a continuous and complete spectrum like solids.

With gases it is very different. It has been pointed out on page 13 that in the air the molecules have an average free path between the impacts something like two hundred times their own diameter. They have, therefore, time to vibrate at their own periodic rate *without any interference* a great many times a second. Thus, if a molecule of hydrogen when vibrating gives out red light having a wave-length of one forty-thousandth of an inch, it shows that it is vibrating as many times a second as one forty-thousandth of an inch is contained in 186,000 miles. This is

$$\frac{186{,}000 \times 5280 \times 12}{\frac{1}{40{,}000}} = 450 \text{ millions of millions.}$$

It collides with others nearly 20,000 millions of times per second, yet between any two impacts it has time to make

$$\frac{450 \text{ millions of millions}}{20{,}000 \text{ millions}} = 22{,}500 \text{ vibrations}$$

without being interfered with for every one that is in any way hindered, so the waves are nearly all of a uniform length. The prism refracts them all alike, and thus produces a spectrum characteristic of the substance and also of its condition as a gas. An incandescent solid gives a continuous spectrum, and incandescent gas gives a discontinuous bright line spectrum. In the experiments described with the solutions of salts giving colored flames, it is to be remembered that in the flame the solutions are at once converted to gas, and the molecules have the free path needed for giving their proper spectra.

If the tip of the lower carbon of an arc lamp have a cavity made in it for holding in turn small bits of different metals such as copper, zinc, and silver, when the upper carbon touches it the metal is not only fused but is vaporized; its high temperature causes it to give out light with its characteristic wave-length and color. And these may be thrown upon the screen with the apparatus shown in Fig. 127.

Absorption Power of Gases. — Produce a good continuous spectrum with either sunlight or an arc light as described on page 235. Then with a gas flame vaporize a lump of sodium as large as a pea in an iron spoon, and hold it in the path of the rays between the slot and the prism. The spectrum upon the wall will now be seen to have a well-defined and very black line across the yellow part, due to the fact that the gaseous molecules of the elements are able to absorb and stop such waves as they are themselves able to produce. These are therefore abstracted from the beam, while the remainder go on and form the rest of the spectrum. The black line thus produced is called an *absorption line*, and is as much characteristic of the element sodium as is the bright yellow light which itself can produce. It indicates more than this, for it shows the existence of a source of light behind it which is either solid or liquid, and that itself is in the gaseous form, else it could not absorb the particular rays it does.

By making the slot of the spectroscope quite narrow and examining sunlight with it, the spectrum is seen to be crossed by a large number of fine black lines parallel to each other. They may be shown on the wall by making the slot in the apparatus (Fig. 125) narrow, and nicely focusing it before placing the prism in its path. One may observe a line in the yellow, several in the green and in the blue, and some broad ones near the limit at the violet end of the spectrum; and with a lens having a focal length of four or five feet, some hundreds may be seen. These black lines in the solar spectrum are known as *Fraunhofer lines* — named after the one who first studied and made a chart of them.

Their origin is the same as the black sodium line already described as due to absorption in a gaseous medium. The body of the sun that gives out most of the light appears to be an incandescent solid or liquid, and gives out light of all wave-lengths. Its temperature is so high that all substances at its surface have more or less a gaseous form; that is, the atmosphere of the sun is made up of the highly heated gases of the substances that compose the body of the sun. The light from the latter has to travel through a great many miles of this atmosphere, and in it absorption goes on, each element stopping those waves which belong to its period of vibration. When the beam of light reaches the earth it has lost some of the constituents it started with. The Fraunhofer lines indicate which ones have been lost, and measuring the corresponding wave-lengths enables one to discover the elements which are present in the atmosphere of the sun. Nearly every element we are acquainted with on the earth has been identified by the lines seen in the solar spectrum, some elements giving but a few lines, others, such as iron, giving several hundred.

The spectra given by the moon and planets show the same lines as those given by the sun; for the light from them is only sunlight reflected from solid bodies, which do not change the character of the light, only its direction. The spectra of stars show similar constitution and condition to the sun, but differ in the relative amounts of the different elements present.

Comets give gaseous spectra, much hydrocarbon being present; and many cloudy patches in the sky, called *nebulae*, seem to be altogether gaseous with hydrogen in abundance. In the long reaches of space between the stars and the earth there are stray molecules of many kinds which by their absorptive action show their presence in small quantities. Benzine and other alcohol derivatives are particularly noticeable. Thus a beam of light can inform us of the kind and quality of matter from which it originated, whether it was solid or gaseous, and the kind and condition of the matter through which it has passed. The elements that are present in the sun and its condition are as definitely known as it would be if it were no more than a thousand miles away instead of being so many millions, for ether waves do not change their character nor their wave-length by traveling in free space.

Invisible Waves. — We know that sunshine warms the earth, and the hand feels the warmth of the direct rays. When a spectrum is formed of the sunlight, the rays are spread out into a band, as already described on page 235, and if one reflects upon the fact that the waves represent the heat energy of the sun, he might expect to find this energy distributed through the spectrum; but if the hand be held in it, it does not appear to be warm in any part. This is because the amount of light used to produce the spectrum is small,

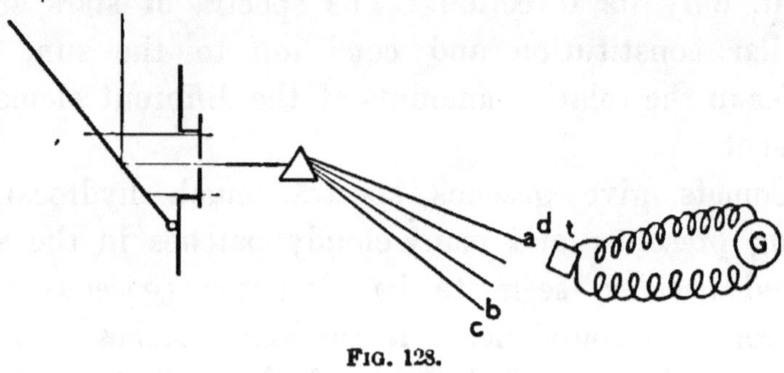

Fig. 128.

and it is so much diffused that the heat is not great enough to be felt. If more delicate means are employed to test it, the heating effect is found in every part of it. For this purpose a delicate thermopile may be taken. When its face is heated it produces a current of electricity, and if the wires are connected to a delicate galvanometer, its needle is moved and indicates the difference in temperature. Let the face of such a thermopile (Fig. 128) be moved through the spectrum not very far from the prism, the galvanometer needle shows great difference in different parts of the spectrum. It indicates but little at the blue end **b**, but shows more and more as it is moved through the green and yellow, and is very much greater at **a** in the red. If it be moved beyond the red towards **d**, where there is no light of any color to be seen, the needle shows still more heat than elsewhere, and proves that there are waves beyond the red that have more energy than any in the visible spectrum, and that they, too, are refracted like the waves that can be seen.

Fig. 129.

A *bolometer* is much more sensitive than a thermopile. It consists of a thin filament of carbon **f** (Fig. 129), which makes a part of an electric circuit with a battery **B** and a galvanometer **G**. The resistance of the circuit is changed by different temperatures of the filament, and the needle of the galvanometer indicates the changes. When this filament is moved through the

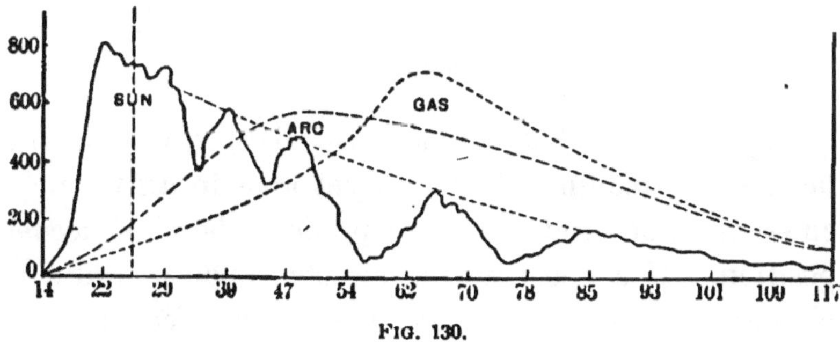

Fig. 130.

spectrum, the presence of the Fraunhofer lines is shown by the falling of the temperature in the filament. With this instrument the distribution of energy in the whole spectrum has been charted (Fig. 130), and it shows a curious lack of uniformity. In this diagram the visible spectrum is that between 0 and the vertical dotted line. All to the right of that line is below the red and therefore of greater wave-length; this part extends to several times the length of the visible part, with a number of places where there are gaps, showing the lack of proportional parts of certain wave-lengths.

The line marked *arc* shows a different but more uniform distribution of energy, the greatest amount being far below the red end. The curve for *gas* has its maximum still further away and rises higher. The relative amount of energy in the visible part is greatest in sunlight, and is at its maximum in the brightest part of the spectrum, the yellow. In the experiment tried as above, the greatest effect will be found below the red; but that is occasioned by the fact that the glass prism absorbs a large amount of the energy of those visible waves and becomes heated.

PHOTOGRAPHY.

Beyond the violet there are still shorter waves, with but a small amount of energy, yet able to affect molecules of some kinds so as to produce chemical decomposition. If a spectrum be directed upon an ordinary photographic plate, the plate will not be affected by the red and yellow rays; by the green it will be affected slightly, but increasingly towards the blue and violet, and beyond these decreasingly to some distance, thus showing that there are waves too short to be seen, and with too little energy to affect coarser apparatus, yet enough to affect molecules and decompose them, for that is what happens in photography. The material in common use for making a surface sensitive to light is the bromide or the iodide of silver—substances that are not very stable compounds; the light of these short wave-lengths dislocates their atoms. The solution into which the plate is put after exposure washes out the bromine or the iodine and leaves the silver in a deposit, which is thicker or thinner as the action of the light has been greater or less. By mixing with the silver solution such substances as the anilines, the whole is affected by longer waves, and in this way the whole spectrum has been photographed below the red end as well as all the visible spectrum. Photographic action is the changing of molecular structure by means of ether waves, and different kinds of molecules are most affected by different wave-lengths. It happens that silver salts are more sensitive to the waves called blue, but what are called *blue prints* are made with salts of iron.

In nature the photographic action of sunlight is to be seen in every direction. The darkening of shingles and clapboards on houses, the fading of the colors of fabrics and of paints, the tanning of the skin exposed to the light, the coloring of flowers and of fruit, which will not take place if kept in darkness, are all the result of molecular changes brought about by the action of ether waves; not any particular kind of waves for all, but each substance is more affected by some kinds than by others.

A thick paper star or other design if pinned to the freshly planed surface of pine or white wood, or pasted to the side of a green apple and left, will have its outline photographed after exposure to sunlight.

I. Electric Photography.—In ordinary photography the sensitized surface is acted upon by ordinary light waves such as can affect the eye, but waves originated by electric action and much longer are competent to do the same thing in the dark. Paste a piece of tin foil, as large as the hand, upon a sheet of glass twice as large, and connect the foil by a wire to one of the discharging rods of a Holtz or Wimshurst electric machine. Another glass prepared in similar way and connected to the other rod of the machine may be laid upon the first, the two glass sides together. When the machine is worked and sparks pass between the knobs, electric waves will pass between the sheets of tin foil through the glass, and if a sensitive photographic plate in its holder be placed between the glass plates, it will be acted upon as if in the light. A coin placed upon the holder will have its image impressed upon the sensitized plate, and it may be developed in the common way. Such pictures are called *electrographs*.

II. X-Ray Pictures. — When a highly exhausted Crookes' tube is lighted up by electrical discharges from a glass-plate machine or induction coil, the character of the discharge is seen to be unlike at the two inner terminals. The wire terminal, which leads the electricity into the tube, is called the *anode*, and the one leading out of it, the *cathode;* electrical discharges from the latter produce upon the inner surface of the glass the greenish phosphorescence, which can be seen, and upon the outer surface invisible waves which pass freely through wood, the fleshy part of the hand, or other parts of the body, and still produce phosphorescence.

FIG. 131.

By using a screen coated with phosphorescent material, such as the tungstate of calcium or platinocyanide of barium, inclosed in a dark box (Fig. 131), one may see in these radiations from the tube the coin in a closed pocketbook, the shadows of the bones in the hand, and pieces of metal imbedded in the flesh. The same radiations act upon a photographic plate inclosed in its holder if held near the tube. As these radiations are not reflected, refracted, nor polarized like ordinary rays of light such as affect the eye, they have been called *X-rays*, to indicate an unfamiliar form of radiant energy.

MECHANICAL EFFECTS OF ETHER WAVES.

The Radiometer. — It has been stated that whenever ether waves of any length fall upon matter of any kind, some of it is reflected, else the body could not be seen, and some of it is absorbed by the body. Such as is absorbed is at once transformed into heat and raises the temperature of the substance. A surface painted black absorbs nearly all the radiant energy that falls upon it, and so may become hot in sunlight. A body with a surface hotter than the air adjacent to it is continually heating the air. The molecules of air that strike upon the body are beaten back with more energy than they had before striking, and their rebound reacts upon the surface producing a pressure upon it. This increased pressure is made apparent by the device called the *radiometer*. It consists of a small mill made with disks of mica **a b c d** (Fig. 132) blackened on one side and fastened to four arms mounted so as to spin upon a needle point, the whole inclosed in a glass bulb from which a large part of the air has been drawn. When radiant energy falls on the blackened surface of **a**, the surface is heated more than when an equal amount falls on an equal surface of mica **c** not so blackened. The molecules of air will bound away with greater velocity and, therefore, with more energy than they had when they struck that surface, and will produce a pressure tending to move the disk backwards in the direction of the arrow. If the mill wheel is delicately poised and light enough, it will turn until the disk on **b** is in the position in which **a** is represented to be, and so keep up a continuous rotation. If the air be of common density instead of being reduced, the free path would be a short one, and the quickened motion of the molecule would be handed over to the one next it, and so conducted away even to the other side of the disk before the disk would have time to move. With less density and longer free paths, this distribution of gaseous pressure cannot go on so fast, and the disks move. It is not the impact of ether waves upon the disks that produces the pressure; the radiant energy is transformed into heat — molecular vibrations — and this into the mechanical pressure of free-moving gaseous molecules.

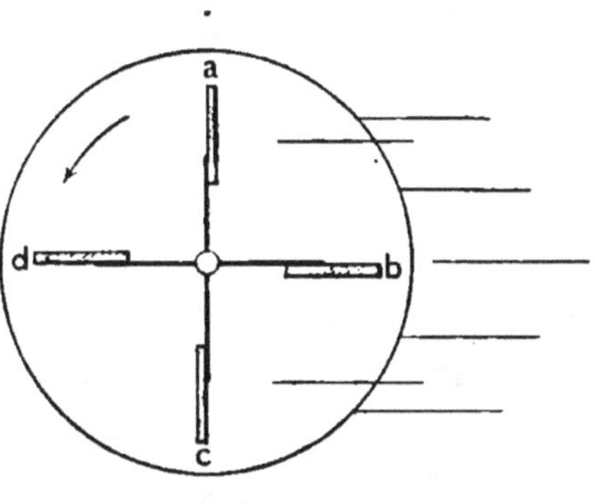

Fig. 132.

A delicate radiometer will show by the rate of its rotation the difference in the distribution of the energy in the spectrum better than the thermopile and galvanometer show it, as represented in Fig. 128.

Double Refraction. — When a clear crystal of Iceland spar is laid on a printed page, every letter seen through it appears double, and the distance apart of the images depends upon the thickness of the crystal. If the crystal be twisted round, one of these appears to revolve about the other. This may be seen to great advantage by directing a small beam of sunlight through

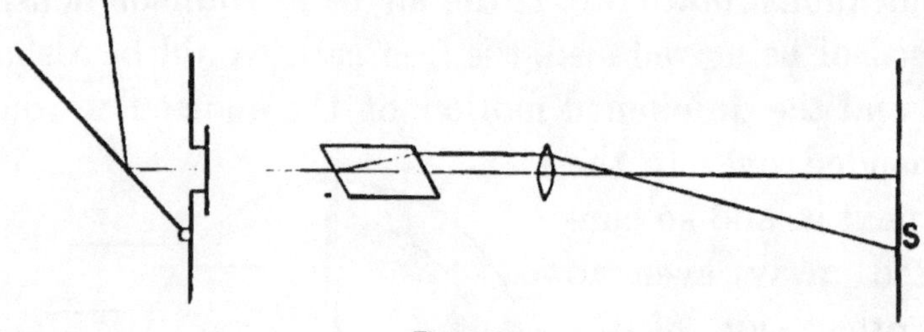

FIG. 133.

the crystal, and then with a lens projecting the two images that will be formed of the hole, as in the diagram. By rotating the spar on the beam as an axis, one of these images will go about the other. This means that the light that produces the second image always goes through the spar in a certain direction that depends upon the molecular arrangement in the spar itself. On examining the spar it will be noticed that it has acute and obtuse angles, and each face is a rhomb (Fig. 135). When the light falls upon one of these faces, a part of it goes through nearly in a straight path **a b**; the other is refracted *towards the obtuse angle* and emerges at a distance from the first, and then goes on *parallel* with it as shown in Fig. 134. The ray **a c** that is most refracted is called the *ordinary ray*, for it follows

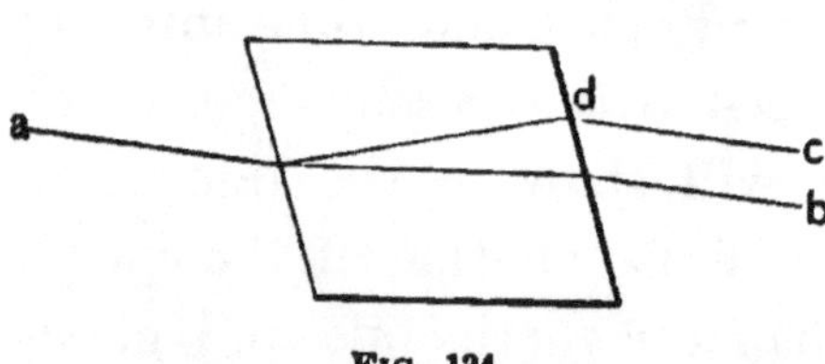

FIG. 134.

the common law of refraction the same as for glass. The other ray **a b** does not follow that law, and is for that reason called the *extraordinary ray.* This phenomenon is called *double refraction.* If common white light be used for these experiments, both these images appear alike in brightness, but they both show another property not possessed by common white light. Let a second piece of spar similar to the first be placed between it and the lens in Fig. 133, so the light of both beams will fall upon it, each of these two beams will again be divided ; now if one of the spars be rotated as at first, two of the beams will rotate about the others, but at certain positions one pair of them will disappear, and at right angles to this the other pair will disappear, while the former ones will be bright, showing that whether the light gets through the second prism depends upon its position.

Polarized Light. — Let a piece of glass, the size of the glass in the *porte-lumière* be painted black on one side so as to absorb all the light that goes through the glass to it ; fix it over the reflector of the *porte-lumière* so as to be used as a reflector in its stead. The light reflected from this blackened glass will be seen to be much less than that from the silvered surface of the common mirror. Direct a small beam of this light through the spar as before, and rotate the spar. When the shorter axis **A B** of the face of the spar is parallel

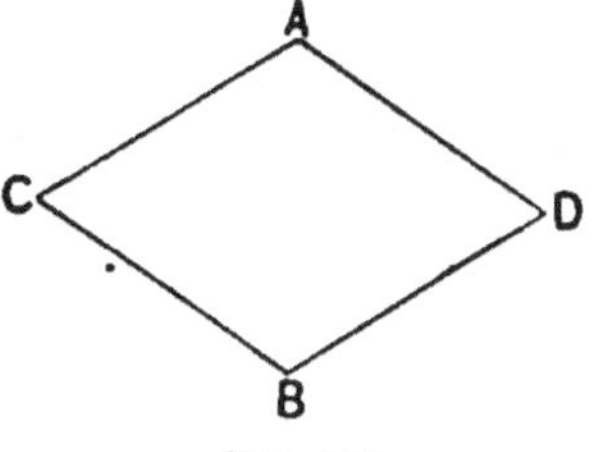

FIG. 135.

with the inclination of the blackened glass reflector, the light will go through it; when the spar has been turned through 90° and **C D** is parallel with the inclined mirror, the light will be quite shut out, as was the case in the

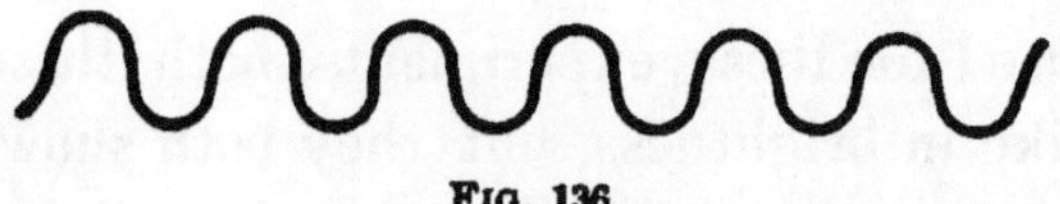

Fig. 136.

use of the second spar in the former experiment. This phenomenon may be understood by bearing in mind that light consists of wave motions. A model of such waves may be made of wire bent thus. If one looks along such a model in the direction of its length, he will see only a short straight line. All the waves will be in one plane, and of course could go through a structure as between the fingers, if it were moved in the plane of the fingers. Turned at right angles to the fingers, it would not pass through. A beam of light in which all the waves are in the same or parallel planes is called a *plane polarized beam*, and anything that causes the waves to vibrate in this way is called a *polarizer.* Such light as is reflected from this front surface of the blackened glass consists of waves which are *vibrating in planes parallel with its surface* as **a b** (Fig. 137). The light which is vibrating in a plane perpendicular to

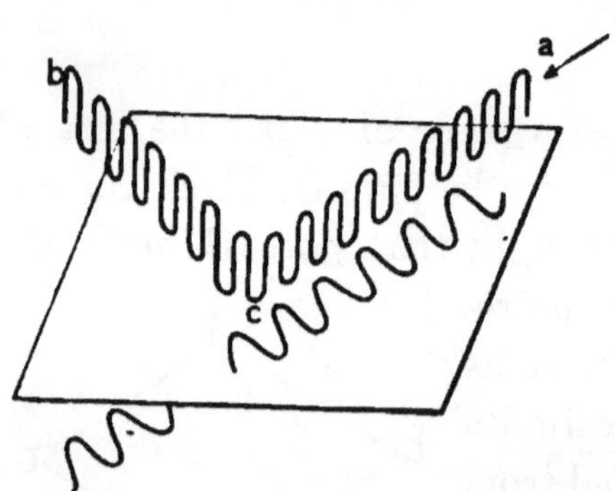

Fig. 137.

the surface **c** goes through the glass and may be absorbed by the black back.[1]

Hence the reflected light consists of light waves that are vibrating in parallel planes and is, therefore, plane polarized. The spar permits light to go through it in only two planes, **a b** and **c d** (Fig. 134). When, therefore, the light reflected from the glass comes to the spar, whether it gets through it or not depends upon whether one or the other of the spar's planes coincides with the plane of the polarized light. A thin section of a piece of tourmaline (Fig. 138) possesses similar qualities — permitting only such rays to go through it as are vibrating in planes parallel with its length. It is, therefore, a polarizer, and two such pieces crossed upon each other stop all light, though apparently both are transparent.

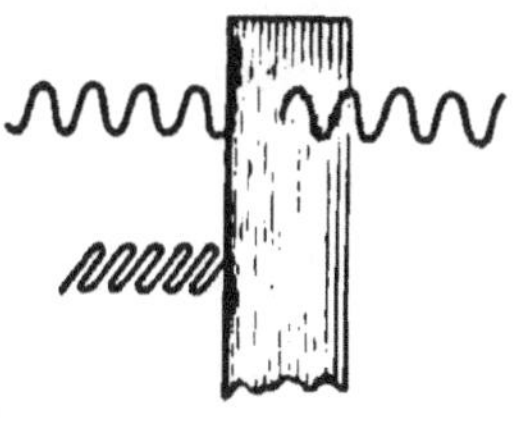

FIG. 138.

The Nicol's Prism. — If a long piece of Iceland spar be cut in two in the line **a b** (Fig. 139), its short axis, and then cemented together again in the same position, the extraordinary ray **c d** goes through without trouble, while the ordinary ray which meets the cut surface at a

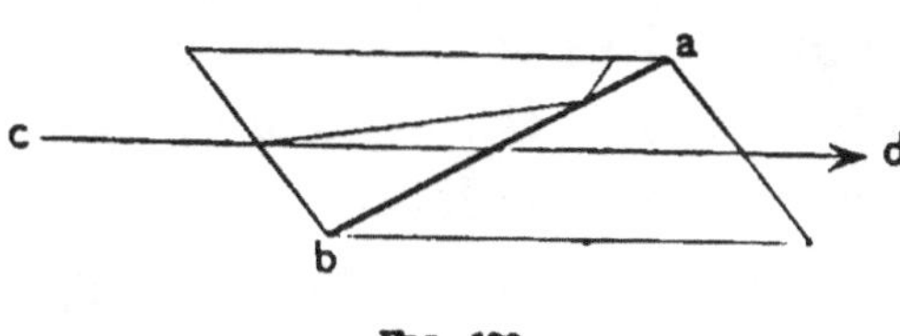

FIG. 139.

[1] Some scientists consider the plane of polarization to be at right angles to the plane of vibration. In this book the plane of polarization is considered the same as the plane of vibration.

smaller angle is totally reflected and goes out on the side near **a**. This leaves the beam at **d** composed of rays polarized in the plane of the shorter axis of the face **A B** (Fig. 135). This is called a *Nicol's prism*, and is of great service in the study of polarized light. With this one may test the character of the light reflected from any surface or from any source. Reflected light from most surfaces, as from the floor, table, walls, or the sky, is thus discovered to be more or less polarized, for by rotating the prism the light appears to be more

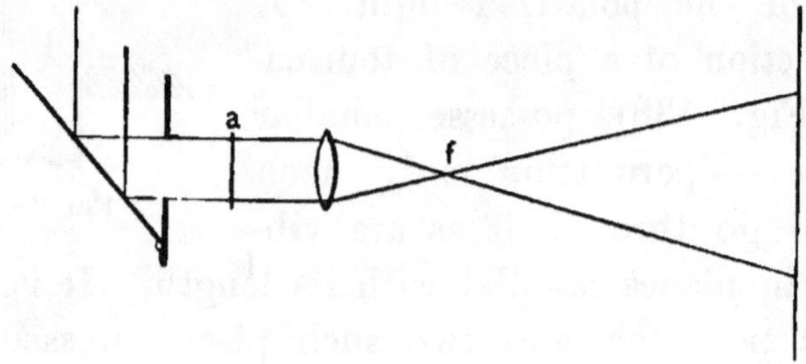

FIG. 140.

or less bright according to the angle through which the prism is turned. When used in this way to discover the existence of polarized light and its plane, it is called an *analyzer*. Let a large beam of plane polarized light be directed through the lens so as to give a disk of light upon the screen (Fig. 140). If the Nicol's prism be put in the focus at **f**, so that all the light goes through it, it may be rotated so that all the light will be stopped and the screen be dark. Let a sheet of mica be placed at **a**, the light at once goes through the prism; on the screen it appears of some color which depends upon the thickness of the mica. If this be of different thickness, each part will have a different tint; and if a geometric diagram be cut in it having different thicknesses, the diagram will show in bright colors. Rotating the mica in its own place will change these tints. Other thin crystals will show similar effects, especially selenite, and microscopic crystals, if projected in the ordinary way as described on page 118, often will show in this light very beautiful tints, and may be identified in this way.

These results are due to the fact that such crystals are double refracting and break up each ray into two parts having different directions; when placed in polarized light from some other source, interference is pro-

FIG. 141.

duced and some of the elements of white light are cancelled, leaving the rest of the constituents. How this can be may be understood by taking two wire models of waves of equal length, as **a** and **b**. If one of these as **a** be moved along **b** until the crests of its waves are over the troughs of **b**, the two will represent the conditions when they will necessarily cancel each other. This takes place when one is half a wave-length behind the other. The waves of light are so short that only a thin section of a crystal is needed to effect this. Glass that is not well annealed shows colors when examined in polarized light, for the molecules are in a state of stress within it, and are not in stable positions: double refraction takes place in such places, and the directions of the stress are shown by the colors; they vary with the form of the glass. Annealed glass which shows no colors will show them plainly if bent or twisted or compressed.

Diffraction. — If light from the *porte-lumière* be directed through a small hole into a room which is otherwise dark, the beam may be traced straight across the room to the wall or screen by the dust in the air, but the orifice itself may be seen from every part of the room, which shows that light is to some extent deflected at that point. If the first orifice be a round hole a quarter of an inch in diameter, and another similar hole be cut in a sheet of paper **A** and placed in the line of the beam, another sheet of white paper **B** placed a few inches beyond will show a series of colored circles upon it, the inner one bluish, and the others following the same order as that in the solar spectrum. This phenomenon is called *diffraction*, and it shows that ether waves like other kinds set up waves

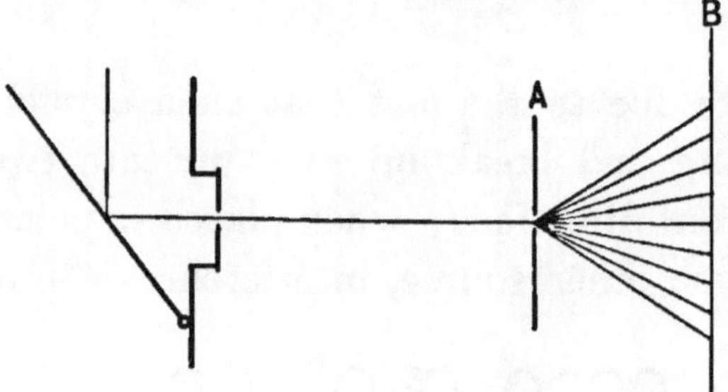

FIG. 142.

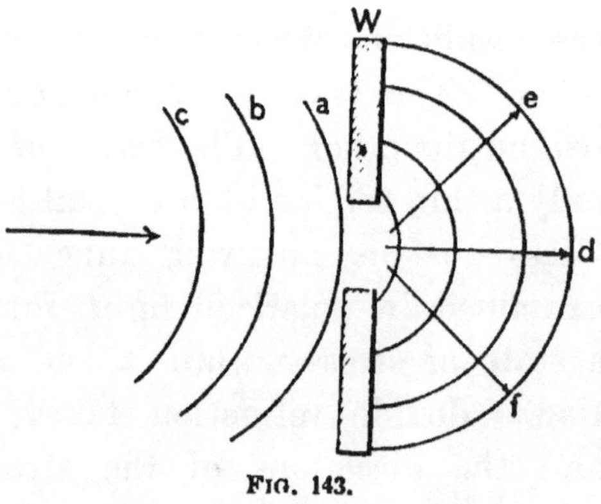

FIG. 143.

in a lateral direction. Let **a b c** represent waves of water moving in the direction of the arrow. When **a** reaches the wall **W**, with an opening in it so a portion of the waves can go through, that portion will not only continue on towards **d**, but will spread out to **e** and **f**, and so will each succeeding wave.

If fine lines are ruled upon a piece of glass, five hundred or more to the inch, and a beam of sunlight be sent through it, very beautiful spectra can be produced, showing the Fraunhofer lines. This is called *diffraction grating.* The finer the rulings of these lines, the longer is the spectrum. The colors are produced by interference. Thus, let **A** and **B** (Fig. 144) represent the clear space between the lines through which the rays to **C** and **D** can go straight forward. At **A** the defracted waves start towards **D**, **E**, and **F**, where they meet with other waves from **B**. It is plain that ray **A D** has traveled a greater distance at **D** than has **B D**, and **A E** than **B E**. Whenever this difference in the path of the rays amounts to half a wave-length, the interference cancels them, but **A D** would be shorter than **A F**, and would represent a shorter wave. The colors then begin with the shortest visible waves — the blue — and continue through the spectrum in the wave-length order. These phenomena may be made very bright and plain by employing the

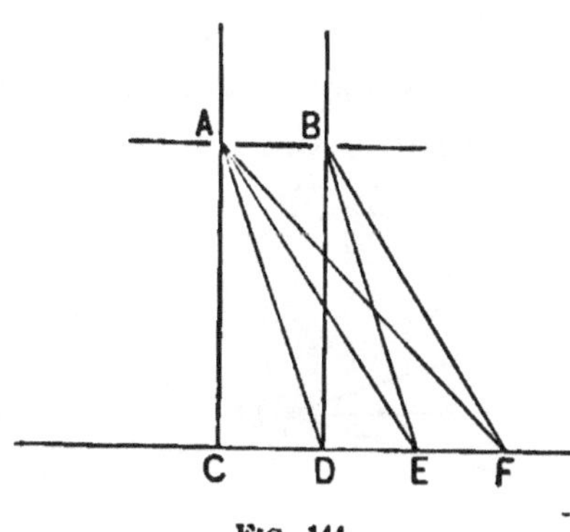

FIG. 144.

porte-lumière and a slot **A** (Fig. 145), the twentieth of an inch broad, like that used in the spectroscope, and a common lens **O** for sharply focusing the image of it upon the screen **S**, and the grating placed at the principal focus in front at **D**. It is with spectra produced in this way, with grating ruled 43,000 lines to the inch, that our knowledge of the sun and stars has been greatly extended. Such gratings are able to produce spectra forty feet long, and a great amount of detail may not only be seen, but photographed.

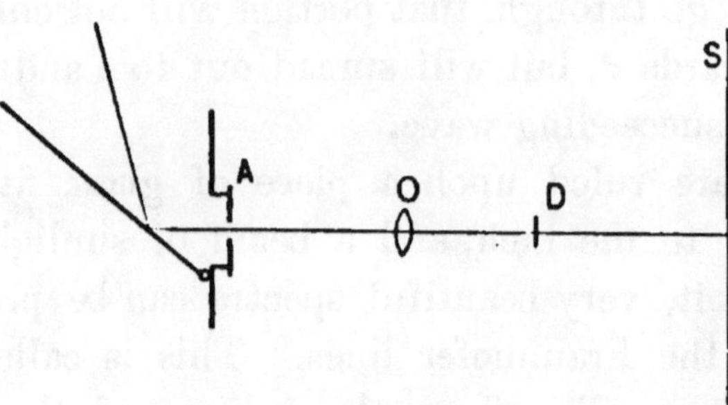

FIG. 145.

The Eye and Vision.—The physical structure of the eye is in many particulars quite like the photographic camera, the latter, of course, being simpler in structure.

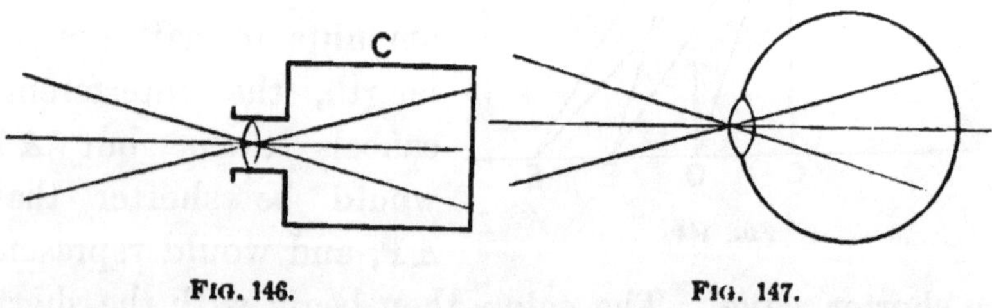

FIG. 146. FIG. 147.

The camera **C** (Fig. 146) consists of a box having an adjustable lens in front for producing an image upon the sensitive surface placed at the back. In like manner the eye is a globular chamber (Fig. 147) having an adjustable lens in front for producing an image upon a sensitive surface at the back. In order to allow more or less light to pass through the camera lens, strips of metal called *diaphragms*, having holes of different sizes in them, are placed in front of the lens. In the eye, to effect the same thing, a muscle called the *iris* (Fig. 148, i, i) is fixed so its contractions in different degrees change the size of the pupil of the eye, the pupil **E** being the orifice in this muscle. The iris is of different colors in different individuals, being blue, gray, brown, or black. A ratchet wheel in the camera enables one to focus sharply images upon the back. In the eye the lens **F** is itself provided with muscles at its edge, which by contracting make the convexity greater or less, and so change the focus.

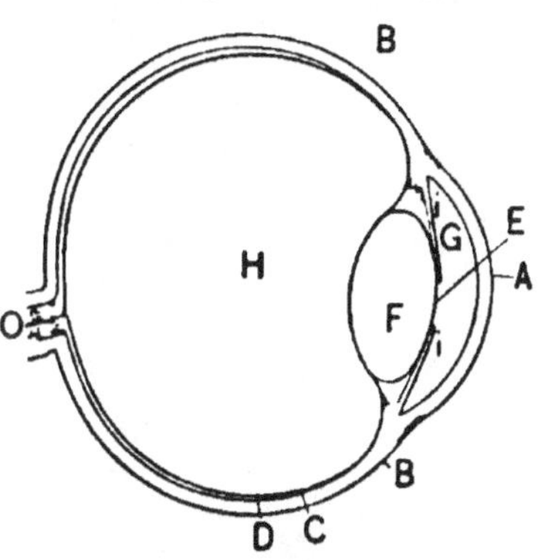

FIG. 148.

In the camera the sensitive surface upon which the picture is made is changed by the removal of the prepared plate. In the eye the surface is continually renewed by the physiological process called *secretion*. In the camera the sensitive substances upon the plate is a preparation of a salt of silver. In the eye the sensitive substance is a complex chemical compound called *purpurine*, secreted by the retina **D**, which is the name of the surface at the back of the eye.

The globular chamber of the eye **H** is filled with a thin jelly-like liquid called the *vitreous humor*. The lens is a transparent elastic body made up of layers somewhat like an onion; its change of form can be produced by the conscious effort to produce distinct vision, which is called *adjustment*. In the normal eye the focus of the lens is at the retina, but in some individuals the curvature of the lens is too great or too small, and distinct vision is impossible without aids. If the lens is too much curved, the focus is too short, and a concave glass is needed to lengthen it. In the other case the focus of the lens is beyond the retina, and it must be shortened by proper convex lens.

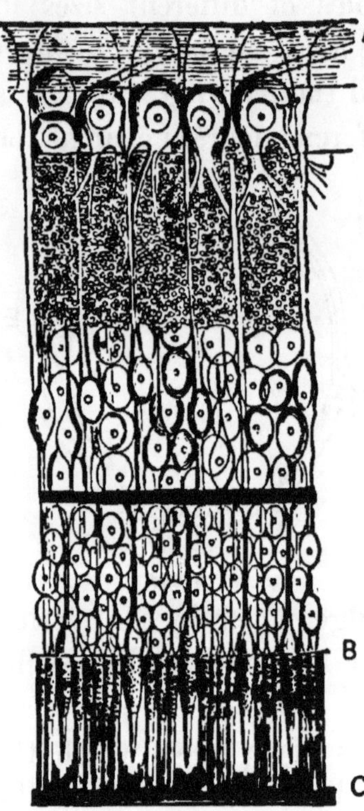

FIG. 149.

A, the inner layer of the retina, next to the vitreous humor.
B, the layer of rods and cones.
C, the black layer which stops all rays from going beyond. The light reaches the rods and cones after it has traversed all the cell layers between **A** and **B**.

A cross-section of the retina (Fig. 149), when examined with a microscope, is seen to be a complicated structure made up of a number of distinct layers of cells, rods, and cones. These are connected with the nerves which are spread throughout and gathered together in a bundle at the back of the eye O (Fig. 148); this is called the *optic nerve*, and goes to the base of the brain. Its function is to transmit retinal disturbances to that seat of sensation in the brain called the *sensorium*, where they are interpreted, and we are then able to say we see.

Phenomena of Vision. — The image formed upon the retina is inverted for the same reason that images formed by lenses in other places are inverted. How it is we see things upright has not yet been explained. The range of ether wave-lengths capable of affecting the retina is relatively small when compared with all the waves present (Fig. 130), but the amount of energy needed to produce vision is almost incredibly small. The energy of a wave depends upon its amplitude; how small that must be in waves that come from stars millions of millions of miles away! Their wave-length does not change, but their energy is inversely as the square of the distance they have traveled. The eye is more sensitive than any photographic preparation yet discovered.

It is a physical disturbance that produces the sensation of light, and this may be effected in a number of ways without these waves. Thus, with the eyes shut, let one press with the finger upon the eye and a circle of light can plainly be seen. A sudden bump upon the head or an electric discharge through the body produces a flash, and this shows that *light* is not a thing outside the eye, but produced by a physiological effect in the eye.

Persistence of Vision. — When a firebrand is swung round there appears to be a trail of fire the length of which depends upon how fast the brand moves. If it be swung round at the rate of about ten revolutions a second, the trail makes a complete circle. This shows that the sensation of the light does not cease the instant the source is removed, but lasts about the tenth of a second for light not very bright, and it may last much longer for any bright light. If the sun be glanced at, one may see the image of the sun for some seconds, even on closing the eyes. If a well-lighted window be steadily looked at for a few seconds, and the eyes then be turned to a dimly lighted wall, the window frame may be plainly seen. This implies that after action the retina requires a short time to replace by secretion the used-up material. Until that be done completely, there will be parts of the retina less sensitive than the rest, and these parts will give a visual image less bright than the remainder.

COLOR SENSATION.

Complementary Colors. — With the *porte-lumière* and apparatus (p. 229), project a disk of light 3 or 4 feet in diameter upon the screen. Insert a piece of red glass in the beam so as to make the disk red. Now look steadily at the middle of the disk for five seconds; the red glass may then be removed while the eyes are still looking at the same point upon the screen. The disk will now look decidedly green. If green glass be substituted for the red and the experiment tried again, the disk will appear red. Blue glass will give an after effect of yellow, and yellow will give one of blue. If half the disk be colored with one glass, and the other half with some other, when the glass is removed the two halves will show different colors. The two colors that stand related to each other in this way are *complementary colors.* This shows not only persistence of vision as in the former case, but also that a portion of the retinal sensitive substance may be affected by light of certain wave-lengths, while the remainder is not affected, and is, therefore, fresh for use when white light acts upon it. Such phenomena have led to the belief that the photographic material of the eye is composed of three or four constituents, one of them sensitive to red waves, another to the green, the third to blue, and possibly one to degrees of black and white. If red rays alone act upon the eye, the other components will not be affected. When the red material is used up and white light is again presented, the yellow, green, and blue rays give together a sensation of green, for yellow light and blue light when mixed produce whiteness, as can be seen by combining the colors of the spectrum with a mirror so that one color can lap upon another; this leaves the green as the complementary color to red. In general, a complementary color is one which, mixed with another, will produce white light. From such phenomena as these, it appears that vision depends upon the energy acting upon the physiological structure of the eye, what we call light being such ether waves, set up by vibrating atoms, as can affect the sensitized surface of the retina. Similar waves that are longer or shorter than these do not affect our eyes. It is probable that some other living things, such as rats, mice, bats, and insects of various kinds, have eyes sensitive to other wave-lengths, for they appear to see plainly when all is darkness to us. If our eyes were affected by waves of all lengths, there would be no such condition as darkness, for as has before been pointed out (p. 206), all bodies are always radiating waves of some length. Only at absolute zero could there be entire absence of ether waves.

Phosphorescence. — Our ordinary sources of light, whether the sun or artificial, like fire or incandescence of electric lights, are produced by high temperatures, the sun being an exceedingly hot body, as is an electric arc. When an ordinary body like a cannon ball is heated it begins to shine, that is, gives out waves we can see, at about 1000°. These waves are about the forty-thousandth of an inch long and produce the sensation of redness. As the ball is heated hotter, the waves become shorter and shorter until all the rays we can see are produced. There are bodies that give out such waves at ordinary temperatures. Such is the light from decaying wood and fish, from the surface of the sea when stirred by a passing vessel or by strong wind; clouds are sometimes luminous, and the streamers from meteors, at the height of many miles where the temperature must be below zero, sometimes last for several minutes. Fireflies and glowworms show that high temperature is not essential for the production of light waves. They also give out X-rays like Crookes' tubes. Such luminosity of bodies at low temperatures is called *phosphorescence*. It can be produced artificially in several ways. A match scratched in a dark room leaves a luminous streak. Certain kinds of crystals, such as the diamond and ruby, give out light only if rubbed. The sulphides of calcium and barium continue to shine after exposure to a strong light for a few minutes, and a kind of paint has been made of the former which shines for hours in the dark. Electric discharges make glass and gases luminous, as spoken of on page 183. Such phenomena have encouraged the hope that light for commercial and household purposes may yet be produced without the waste of energy in the shape of long waves, which make the greater part of radiations of hot bodies. If all the rays from an incandescent electric lamp were of the visual length, the lamp would give out twenty times more light.

Fluorescence. — If a little quinine be dissolved in a test tube of water to which a drop of sulphuric acid has been added, the solution will appear of a bright blue color. If it be examined in the colors of the spectrum, it will not show this color in either the red, the yellow, or the green; in the blue it is distinct, and will continue to be blue when moved into the region beyond any of the visible rays. This means that the quinine possesses the property of changing the wave-lengths to longer ones. Many bodies possess this quality in some degree, but a few have it in a remarkable degree. Eosine, uranine, thaline are prepared from coal tar. A minute quantity of either of these dropped upon the surface of water in a tumbler will be seen to sink slowly, leaving a beautiful, deeply colored thread from the surface, and will tinge all the water. If this water be examined in the spectrum as before, the particular colors will be found well developed in rays of much shorter wave-length than those of the color they show. Their molecules in some way act so as to increase the wave-length of such rays as are reflected by them. Uranium glass which has a yellowish green tint is thus fluorescent, and shows its colors well in blue light; also when lighted up by an electric discharge. For this reason that kind of glass is sometimes made into vases or other ornamental forms for experiments with electrical discharges, as in Geissler's tubes. Substances that, like these, change the wave-length of the rays that fall upon them into any other length longer or shorter are called *fluorescent* bodies. The above-mentioned substances all lengthen the waves, but naphthaline red in red light reflects yellow light, and chlorophane changes rays below the red of the spectrum into emerald green light. It is a remarkable fact that light that has once passed through a fluorescent solution will not affect another solution of the same kind. All fluorescent bodies are likewise phosphorescent.

QUESTIONS.

1. If light travels 186,300 miles in a second, how long will it take for it to come from the sun to the earth?

2. If it takes $3\frac{1}{2}$ years for light to come to us from the nearest fixed star, how far away is it?

3. How long would it take light to go from Boston to Chicago?

4. Suppose a straight bar-magnet like a compass needle rotates on its pivot once a second; how long will its ether wave be?

5. If it should spin a thousand times a second, how long will its wave be?

6. Two lights give shadows of equal intensity on the photometer, — one is one foot distant, the other is five feet distant from the surface; how much brighter is one than the other?

7. A lens has a focal length of 10 inches; if an object be placed in its axis 15 inches from it, how far on the other side of the lens will the image be? $\left(\frac{1}{p}+\frac{1}{p'}=\frac{1}{f}\right)$

8. If the screen (Fig. 117) be 20 feet from the lens of 1-foot focus, how far must an object be from the lens to produce a proper image?

9. How large would be the image of a nickel coin under the above conditions; that is, how much would be the magnification?

10. If the magnifying power of a lens used as above is known to be 250 diameters, what will be the length of an animalcule whose image is found to be four inches long?

CHAPTER XI.

SOUND.

THE word "sound" is sometimes used to mean the sensation produced through the ears or organs of hearing, and sometimes we mean by sound the physical disturbance in the air that may produce the sensation if it reaches the ear. The former is called the physiological definition, the latter the physical definition, which is the one to be considered here. By listening attentively one may, in almost any place, hear a number of different sounds; as those of a ticking clock, of buzzing flies, the wind, bells, or whistles. We learn to distinguish their sources and direction, but need to call to our assistance some of our other senses, such as sight and touch, in order to understand what takes place when the sound is made.

When we strike the table with the knuckles or with a pencil, there is heard a sound of short duration; but if a bell or a tuning-fork be struck, the sound continues for some seconds.

Sounding Bodies are Vibrating. — If a sounding bell or a tuning-fork be touched with a piece of paper, a buzzing sound will be heard. If the tuning-fork be a large one, the prongs may be seen swinging to and fro. When this movement is no longer visible, it may be made apparent again by swinging the fork to and fro in a bright light; the prongs will have a fan-shaped appearance, which will be lost when the fork stops sounding. Let the end of the thumb be dampened, and, holding it nearly vertical on a table, let it be moved rapidly forward; the vibration of the thumb will be plainly felt and the sound may be heard. A piece of chalk upon the blackboard or a pencil upon a slate may be made in like manner to produce a sound, and a line of dots will indicate how many times the moving body has touched the surface. In every musical instrument there is some part capable of prolonged vibration, — in the piano and harp, the wires; in the parlor organ, the brass reeds; in the flute and cornet, a column of air. In every case a sounding body has a vibratory movement. Sometimes the separate vibrations can be distinguished, as when the thumb is pushed across the table, but generally they occur so frequently, that is, so many times in a second, that the ear fails to distinguish any interval between them, and such a sound is called a *continuous sound*.

If a wheel having teeth like a cog-wheel in a clock be made to rotate, and the edge of a paper card be held against it, one may determine the number of vibrations made by the card when it makes the lowest continuous sound; it will be necessary to know how fast the wheel is rotating, and how many teeth there are in the wheel.

Pitch. — With the above device one may observe another fact: as the wheel turns faster the sound becomes higher and higher until it is a screech. This change in the sound is called the change in *pitch*. As the card vibrates faster the pitch rises, and as its rapidity diminishes the pitch falls until the individual taps of the card may be heard. By quickly drawing the finger nail across the cover of a cloth-bound book, a sound may be heard; its pitch will depend upon how fast the finger moves. Tuning-forks are made so as to vibrate a certain number of times a second, that is, each one has a definite pitch. The one commonly used for experiment is called a *C* fork, and makes 256 vibrations per second. Pianos and organs are tuned to a fork making 261 vibrations per second as a standard; the middle *C* upon the keyboard being put in unison with the fork, that is, made to have the exact pitch of the fork; the remaining keys are tuned by ear from this. By pitch, then, is meant the number of vibrations per second that a sounding body makes. Thus the pitch of the lowing of a cow may be 150 vibrations per second, a locomotive whistle 470, the chirp of a cricket 3000, the squeak of a bat 5000 vibrations per second.

Energy of Sounds. — If a tuning-fork be struck hard and its stem be pressed upon a table, it will give a sound of a certain pitch and loudness, but in a few seconds it will become very weak, and one will have to listen carefully to hear it at all. In like manner the sound will be very weak if the fork, while it vibrates, be held in the hand instead of upon the table, yet so long as it can be heard its pitch will remain the same. One may sing the syllable *la* in any pitch, softly or loudly. Difference in loudness of sounds is called difference in *intensity*, and is due to the amount of energy spent in producing the air vibrations. When the fork is first struck, the movements of the prongs may be seen, and the sound is strong; the sound becomes weaker as the *amplitude of the vibrations* (see p. 46) becomes less, until both movement and sound cease. The loudness of sound depends upon the amplitude of the vibrations, and the amplitude depends upon the amount of vibratory energy the body has.

Distributors of Sound. — Let a tuning-fork be struck, and the stem be touched to the top of the table three or four times; the sound is much louder while the stem is touching the table than when it is held in the hand. A music box will sound louder when it is placed on such a surface than when held in the hand. Let a pin be stuck into the table and a thread tied to it, the other end of the thread being tied to the music box or the fork; if the box or fork be made to sound, the loudness will be much increased by pulling the thread taut. It will be the same if a wire or a wooden rod be used to connect the vibrating body with the table, which shows that wood, metal, or stretched strings are better conductors of sound than air.

If the ends of a long rod or wire were held by two persons, and one should pull or push upon it, the other would feel the change. When it is remembered that the sounding body is a trembling body, it may be perceived why the trembling motions are transferred or conducted away. The condition for such conduction is that the conductor be elastic, and the higher the degree of elasticity, the faster the sound is conducted away. If the sound is conducted away faster, it is plain that the sounding body must be losing energy faster. With a clock or watch in sight, strike a tuning-fork, and, holding it in the fingers, note how many seconds it can be heard. It may be gently touched to the table once in four or five seconds to increase the strength of the sound. Make the fork sound as before, and hold the stem upon the table all the time; it will not be heard as long.

In the absence of elastic bodies, sound cannot be conducted away at all. The air is an elastic body, and serves to distribute sound in every direction. If a sounding body like a bell be placed in a good vacuum, it cannot be heard; the whole energy of the sound is transformed into heat and raises the temperature of the bell.

Sound Waves.—Suppose the open hand be swung to and fro like the prong of a tuning-fork; as the hand moves forward, the air in front of it is condensed while the air behind it is rarefied. The denser air has greater pressure than the more rarefied air, and this difference in pressure causes movements in the air towards the place where the pressure is less. The rate at which such a disturbance in the air travels is several hundred feet in a second. The hand moves so slowly that the difference in pressure is very slight and is equalized almost instantly. If the hand could move fast enough a perfect vacuum would be formed behind it, and if it could swing to and fro fast enough it would maintain a space nearly free from air on both sides of it. A condensation in the air immediately begins to travel in every direction, as a wave started in water travels in a widening circle. The same is true of a rarefaction in the air. A vibrating body forms a condensation when moving forwards, and a rarefaction on the same side when moving backwards; both move outwards in every direction, and together they constitute a sound wave.

The length of a sound wave is the distance from the middle of one condensation to the middle of the next, and, therefore, represents a complete to-and-fro movement of the vibrating body. The fork vibrates a great many times a second, and a wave-length depends upon how far the first condensation has traveled while the remainder of the wave is being completed. The air being composed of molecules, it does not move in a body, but each molecule moves towards and away from the source of the sound. The molecules are, therefore,

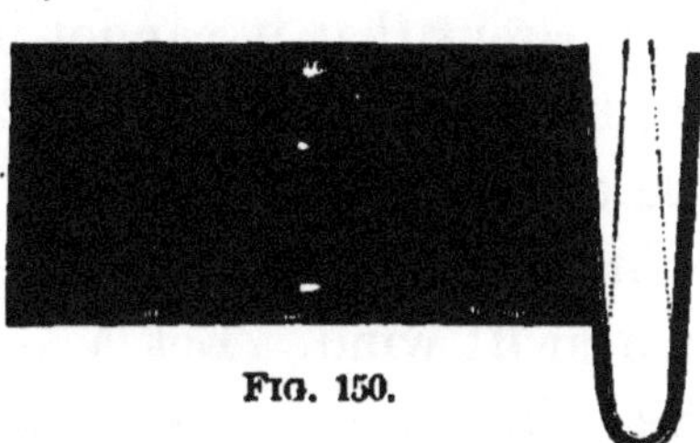

Fig. 150.

. . . .

. . . .

more crowded in some parts of the line than in others. Such a vibration is called a *longitudinal vibration.* As an illustration of this kind of wave motion, take two combs having different numbers of teeth to the inch, for example, one with fifteen and the other with twelve; place the combs together, and, holding them at arm's length towards a window, slowly slide one over the other; series of dark and of light spaces will be seen to move regularly after each other. If the dark space be likened to the dense part of the sound wave, while the light space answers for the rarefied part, the motions will be exactly like those of sound waves. The same thing may be seen often while looking from a distance through a picket fence at another one behind it. If the observer be moving, series of light and dark bands will follow each other like sound waves. Air is so transparent that it cannot be seen, and waves in it cannot be seen, but air waves may be felt. Let one stand one or two hundred feet distant from a cannon when it is fired, the wave of dense air as it passes will be felt as a swift wind. Let a piece of stout paper be tied over the flaring end of a funnel, and the paper be snapped with thumb and finger while the small end is directed towards the face, the puff produced will be plainly felt, and if directed towards a lighted candle, the latter may be puffed out even at the distance of eight or ten feet. The air is condensed within the funnel, and the condensation escapes at once by way of the throat. A series of strokes will give a series of condensations which will follow each other like sound waves.

The amplitude of the vibratory movement of most sounding bodies is very small indeed. A tuning-fork may give a loud sound and still appear quiescent to the eye, the actual displacement of the prong being less than the ten-thousandth of an inch. The movement of the air particles is even less. When it is remembered that sound can be heard in every direction from a sounding body, and that the energy of the waves is rapidly diffused to larger and larger bodies of air, it will be understood that the actual motion of an air molecule when the sound is at the distance at which one can hear another person whistle — say 1000 feet — must be measured in millionths of an inch; for it need be no more than the hundredth of an inch at the starting point, and the intensity varies inversely as the square of the distance from that point. Such a small distance is comparable with the magnitude of the molecules themselves.

Velocity of Sound. — There is a noticeable difference in time between seeing the flash of a distant gun and hearing the report. A bursting rocket and a flash of lightning are followed by sounds after an interval of time, the length of which depends chiefly upon the distance of the disturbance from the observer. Careful experiments have shown that the velocity of sound in air varies with the temperature. At the temperature of freezing water, that is, at 32° F., the velocity is 1090 feet per second, and increases about one foot per second for every degree higher, so that at 70° F. it is $1090 + 38 = 1128$ feet per second. The distance of a flash of lightning may be roughly estimated by allowing five seconds for a mile. It is rather remarkable that thunder, though often very loud, can seldom be heard more than five or six miles, while cannon-firing and powder-mill explosions have been heard fifty or more miles. The reason for this difference is, probably, that the thunder is altogether in the air, while the others are upon the earth and shake it directly.

All kinds of sounds travel with the same velocity. Hence the velocity depends only upon the properties of the air, not upon the source or the pitch of the sound.

Wave-Lengths of Sound. — Suppose a tuning-fork make 100 vibrations per second. At the end of the first second the first wave formed would have traveled a distance equal to the velocity of sound at that temperature, and between the first wave and the last one there would be a line of 100 waves — all alike and of the same length. If the velocity were 1125 feet per second, each wave would be $\frac{1125}{100} = 11.25$ feet long.

Let $v =$ velocity of sound in air,
$n =$ number of vibrations per second,
$l =$ wave-length; then

$$l = \frac{v}{n} \text{ and } v = nl.$$

Given any two of these factors, the other can be calculated.

The velocity of sound is much greater in liquids and solid bodies than in air, for their elasticity is greater.

Sound travels in water	4	times as fast as in air.
" " " brass	10	" " " " "
" " " copper	12	" " " " "
" " " steel	16	" " " " "
" " " oakwood	10	" " " " "
" " " birch	14	" " " " "
" " " pine or spruce	18	" " " " "
" " " glass	16	" " " " "

Thus it appears that sound has a velocity greater than three miles a second in a steel wire, and sound waves that would be ten feet long in air would be 160 feet long in such a wire.

Speaking Tubes. — The sound waves produced by the voice begin at the mouth to scatter in every direction, and every one in a large hall, or at a great distance in an open field, may hear, but the sound is weaker and less distinct at a distance, for the energy of the part that reaches the ear is less. If one speaks into a long tube or pipe, it is not possible for the sound waves to be diffused as they are when in the open air; they retain their energy for a longer time and may be heard at a much greater distance. Conversation even in whispers has been carried on through empty water pipes more than a half mile long. The common speaking tubes used in houses to communicate between distant rooms direct the sound waves without diffusing them.

The String Telephone. — If one will talk or sing into a tin cup, or similar vessel, he will be able to feel the vibratory motions of the bottom of the cup by

FIG. 151. — The String Telephone.

touching it with a finger. The air waves make the bottom swing to and fro as many times as there are waves to act upon it. If two such cups be connected together by a string fastened to the middle of the bottom of each, one may, by holding one cup to the ear, hear easily what another speaks into the other cup, even at the distance of a thousand feet, provided the string be kept taut.

In this case the vibrations of the bottom of the cup give longitudinal vibrations to the string or wire which connects it with the other, and, in turn, the bottom of the second cup is made to vibrate in the same way as the first. The waves are in the string, and therefore are not diffused like air waves, and may be transferred to a greater distance. If the string or wire touch a body like a tree or a house, the sounds will be stopped. If an angle is made in the direction of the string, the latter must be held in position by another string **A** a few inches long (Fig. 152). Let one cup be detached and the end of the string be fastened to a pin stuck in a panel of the door, and a person may hear a whisper made so gently in the cup that no one in the room could hear it at all. This means that the whole door is made to vibrate in the same way that the bottom of the cup vibrates, and that, too, when the sounds are only faint whispers.

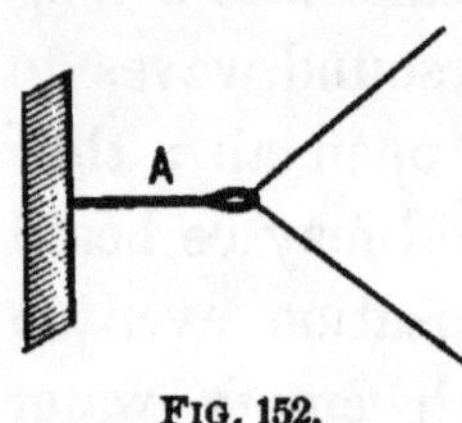

FIG. 152.

In like manner two persons may communicate in whispers through a closed door, one listening with ear against the door, and the other speaking to the opposite side.

With the mouth close to the door or to a wall, the energy of the air vibrations will be mostly spent upon the small surface; at a greater distance the energy will diffuse itself through the room, and the walls, floor, and ceiling will receive only a proportional part, and will be made to vibrate correspondingly. The larger the space, the less will be the energy to the square inch. Whether more or less, it acts in the same way on every elastic surface, and every object in the room, whatever its size, is made to vibrate by every sound. If the walls of the room are thus made to shake, the house is shaken, and the earth in turn has the position of every molecule in it changed in some degree by the energy of sounds. When the vibrations are of molecular magnitudes we call them heat, and sound waves are all ultimately resolved into heat, and are mostly radiated away.

FORCED AND SYMPATHETIC VIBRATIONS.

Every Elastic Body has some Vibratory Rate or Pitch. — When a load of stones is tipped out of a cart, the sound is called a *noise*. It is a kind of roar without any particular pitch; but if each stone be struck with a small hammer, it will be heard to give a sound having a definite pitch. Stones of different sizes and shapes give different pitches, so that it is not difficult to select a series that will give the whole musical scale. Pieces of wood, large nails, bolts, and the like will each give out, if free, a particular sound when struck, as well as an iron poker, a glass tumbler, or a piano-string.

When sound waves in the air fall upon a body which has the same vibratory rate as the waves, the body is not only made to vibrate, but each air wave makes it vibrate a little further, and so its amplitude is increased, just as slight pushes upon one in a swing will cause the swing to go further and further, and continue to swing for some time after the pushing stops. With the voice get the pitch of any key on the piano, then with that key pressed down sing that note to any syllable. On stopping, the piano will be heard to give out the same note, though it has not been struck. The air vibrations have set it in motion. A different key pressed down will give no response. Such an instrument as a bass viol, even if it be many feet distant, may be made to respond loudly by sounding with the voice the pitch of any of the strings. Two tuning-forks of the same pitch act thus on each other; either will cause the other to vibrate by the impact of its air waves. Such forks are called *sympathetic forks;* and vibrations that cause another body having the same pitch to vibrate are called *sympathetic vibrations.*

Press down the damper pedal of a piano so as to free all the strings, and then sing or speak or make any kind of a sound, and the strings will at once respond loudly. There are so many strings with so many different pitches that it makes little difference what pitch is chosen; there will be some strings in unison with it, and they will be set vibrating *sympathetically.*

If a body having a definite rate of vibration is made to vibrate at some different rate, such vibrations are called *forced vibrations.* Thus, when the stem of a vibrating tuning-fork is held upon the table, the latter sounds loudly, but the rate of vibration of the fork is not the same as the pitch of the table, as the table will respond in the same way for a fork of any pitch. The sounding boards of pianos and the shells of violins are thrown into forced vibrations by the sounding strings, and thus the sounds are distributed to a much greater body of air, the strings losing their energy at a swifter rate.

Resonance. — Tuning-forks for experimental work are often mounted on boxes open at one or both ends (Fig. 153). When such forks are struck, they sound

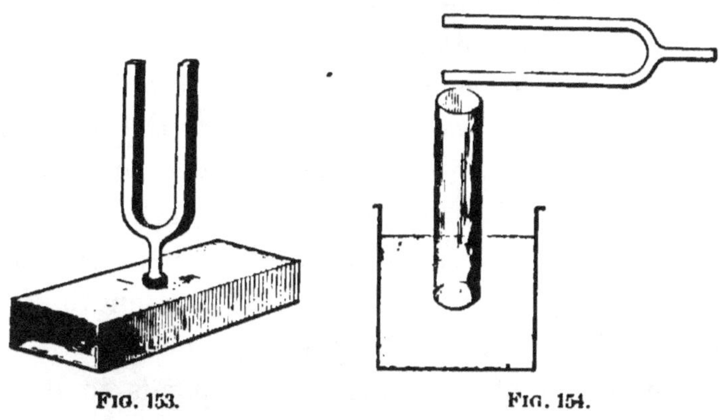

FIG. 153. FIG. 154.

very much louder and can be longer heard; but if the hand be placed over the open end, the sound is greatly lessened. If the fork be removed from the box and the prongs be held in front of the open end of the box, the sound will be nearly as loud as when the fork rested upon the box. Another fork with a different pitch requires a box of a different size to respond in like manner. A pocket tuning-fork, if made to vibrate and held over the end of a tube, as a lamp chimney, may not be heard any more plainly; but if the tube be plunged into water so as to shorten the column of air in it (Fig. 154), some length above the water can readily be found where the sound will be greatly reinforced. The phenomenon is called *resonance*, and such tubes are called *resonant* tubes or boxes. The explanation of this is, the column of air has the same rate of vibration as the fork; that is, it vibrates sympathetically. When the prong of the fork beats downward, forming a condensation, the condensation travels down the tube and is reflected from the bottom to the top; if it gets back to the top at the same time the prong gets to its normal position, the rarefication begins to form at the fork, travels down the tube, and is reflected from the bottom as the condensation was, getting back to the top of the tube when the prong reaches its normal position again; so for each complete vibration of the fork, both parts of the air wave travel twice the length of the tube, — for the complete wave, four times the length of the tube. It follows that the length of the air wave is four times the length of the tube that responds to it. In this way one may determine how many times a fork vibrates in a second: find what length of tube is required for strongest resonance; four times that will be the wave-length l of that sound. Ascertain the temperature in order to know what the velocity v of sound is at that place; then $\frac{v}{l} = n$, the number of vibrations made by the fork. If the tube be a broad-mouthed one, it will be necessary to add to the measured length of the tube about two-thirds of the diameter of the mouth for one-fourth the wave-length. For example, a tube an inch and a half in diameter and twelve inches long was found to respond loudest to a fork. What was the pitch of the fork, the temperature being 70° F.?

$v = 1090 + 38 = 1128$ feet.
$l = 4\ (12 + 1) = 52$ inches $= 4\frac{1}{3}$ feet.
$\frac{v}{l} = \frac{1128}{4\frac{1}{3}} = 260 = n.$ 260 vibrations per second.

In like manner one can determine the velocity of sound in air by knowing how many times a given fork vibrates, and measuring the wave-length in the above way, for $nl = v$.

The resonance of columns of air is the chief source of sound in wind instruments, such as flutes, organ-pipes, and cornets. The holes in the flute and the valves in the cornet regulate the length of the column of resonant air; the pitch of the organ-pipe depends upon its length, and is constant. The mouth is a good resonator for a tuning-fork; by varying the size of the cavity one may find by trial a size that will respond so as to be heard fifty feet away. By snapping the cheek with the finger, the resonance of the mouth may be heard, and a tune may be played by varying its size. For sounds that have a lower pitch, such as those of the voice of a man, which have a wave-length of eight or ten feet, the resonant tube must be correspondingly larger and longer. Places are often found in halls, in long archways, and in some caverns where the voice at a certain pitch causes a very loud response. The amplitude of the air vibrations is made many fold greater by such conditions.

Echo. — Sound waves are reflected from any surface in the same way they are from the bottom of a resonant tube. The tube serves merely to keep the vibrations from spreading in every direction. Such reflected sounds are called *echoes*, and are often heard from buildings or hills not too distant. At the distance of 75 or 100 feet from a building shout "Holloa!" or clap the hands together, and the return sound may be plainly heard. The echo of bells, whistles, and other sounds is common enough, but one seldom

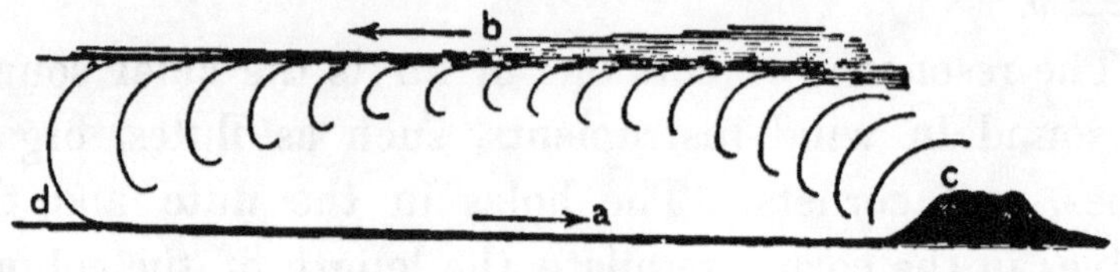

FIG. 155.

or never hears an echo from the clouds. When sound waves rising in the air meet an air surface where the wind is blowing in some different direction from that in which the sound has been moving, they are deflected, and may be directed to the surface of the earth.

This is frequently noticed near coasts where fog-horns are blown as warning signals to vessels. Sounds made at **c** (Fig. 155) are tilted upwards by the wind **a**, so as to leave a space at **a** sometimes a mile or two wide, within which they cannot be heard, though audible at a greater distance **d**.

Vibration of Strings. — If a piece of twine or thread twelve or fifteen inches long be stretched and then plucked with the finger, it will be heard to give out a sound; the tighter it is stretched, the higher the sound it will make. The difference in pitch may be so great for the difference in tension that may be given by the hand that a tune may be plucked on the string by one having a correct ear. The tuning of guitars, violins, and pianos consists in giving the proper tension to the strings.

A **sonometer** (Fig. 156) is an instrument for the study of the relations of the tension, length, size, and weight of strings to their pitch. It is about forty inches long,

FIG. 156.

and generally has two wires stretched between its bridges. These may be tuned like the wires of a piano. There is also a movable bridge, so the vibrating part of either wire can be varied at will. With this instrument one can discover

1. That the number of vibrations a cord or wire will make is inversely as its length.
2. The pitch varies with the tension.
3. " " " inversely as the diameter.
4. " " " " " " density.

In the stringing of a violin or of a piano these laws are observed. For instance, on the violin all four of the strings have the same length; for the highest sounds a string of small diameter is employed, while for the lowest sounds a larger one is used, wound with fine wire to give it greater density, and the intermediate strings vary in diameter. In the piano the wires vary in length, tension, and diameter. The lowest ones are not the longest, but they are much thicker — sometimes are as much as a quarter of an inch in diameter. The upper wires are both short and fine. A good example of the conditions that affect a vibratory body may be shown by stretching a strip of India rubber, and making it give out a sound while its tension is being varied. It will be found to give out the same pitch, whatever the tension, for as it is stretched its diameter is proportionately lessened.

Compound Vibrations.—Let a rope fifteen or twenty feet long be fastened at one end to the wall, while the other end is shaken by the hand. By properly timing the shakes, the rope may be made to swing as a whole, as at **A** in Fig. 157. A faster movement of the hand will cause it to break up into two parts called *segments*, as at **B**; and still faster movement will break it into three segments, as at **C**. In **B** and **C** there are points where there is apparently no movement, as at **n**. These points are called *nodes*. By employing a large tuning-fork in the place of the hand and a long, stout thread, one may observe as many as fifteen or twenty such segments on the thread, the number depending on the tension employed. The experiment shows that a string may vibrate in parts as well as a whole. A stretched wire will vibrate in a similar way. Tune both wires on the sonometer to the same pitch, and place the sliding bridge under one of them so as to divide the wire into two equal parts; each half on being plucked will give the same sound. Place a small paper rider, made by folding a piece of paper two inches long and a quarter of an inch wide, so it will straddle the long wire at its middle. If the short wire be plucked, the rider will remain in place; but if the rider be put elsewhere on the long wire, it will be thrown off. By making several such riders and placing them at different

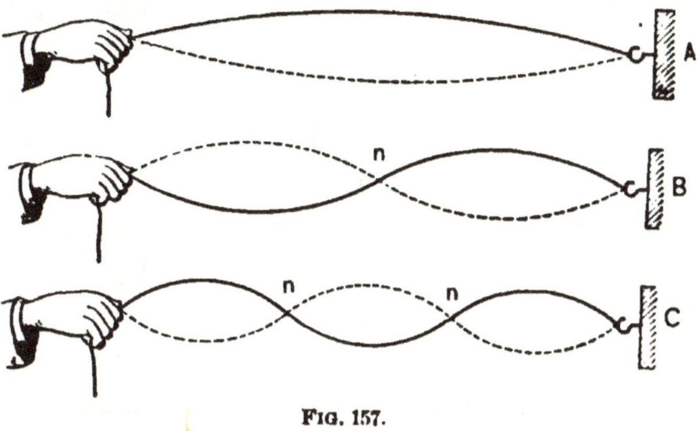

FIG. 157.

places on the long wire, all will jump off except the one at the middle point where there is a node. The long wire vibrates sympathetically with the shorter one, and this necessitates a node at its middle point. If the bridge be moved to one-third the length of the wire, and this wire be plucked, the paper riders on the long wire will indicate by their movements that it has three vibrating parts and two nodes, like **C** (Fig. 157). These vibrations of the long wire will be too minute to be seen, but are sufficiently energetic to toss the papers and thus show how the wire vibrates. Also, one may hear the pitch of the long wire when thus made to vibrate sympathetically with the shortened length of the other wire, but it will be needful to stop the vibrations of the short wire, by touching it with the finger after two or three seconds.

A stretched string or wire may vibrate in more than one period, and produce sounds of more than one pitch. By plucking the wire of a sonometer in different places one may notice the different sounds it gives; some of them may be very high.

One may now notice the different sounds which are given out by many bodies, — a tuning-fork struck with a piece of metal, an iron rod struck with a piece of wood and with a hammer. The different sounds which a violin can give in the hands of an unskillful player are well known, and a piano-string sounds very different when plucked with the finger than when struck by its proper hammer. A piece of stout thread stretched between two tacks a foot or two apart in the window sill, close to the sash, will make many pleasant musical sounds if the sash be raised about an inch so the wind may blow in past the thread. The device is called an *Aeolian harp*. The thread breaks up into many segments, each one producing its own pitch; all are heard together and are highly musical.

The lowest sound that a string or other sounding body makes is called its *fundamental pitch*, and the other sounds are called *overtones* or *harmonics*.

Analysis of Sounds. — With a well-tuned piano one may discover what sounds are present in its longer wires by their sympathetic effects on the upper ones. Thus, gently press down the upper G key of the bass clef (Fig. 158), so as not to strike the hammer; then strongly strike the lower **G**, let it sound a second or two, and then let it up; the upper **G** will be heard sounding loudly, showing that the lower **G** had that component in it. The **D** above the **G** will also be made to vibrate if held in the same way while the lower **G** is struck. In this way one may trace the vibrations of the lower **G** to the keys above (Fig. 159).

FIG. 158.

FIG. 159.

In like manner, by holding down the lower **G** key while the upper keys are struck one at a time, each one will cause the long wire to give out the pitch of the one struck, as illustrated with the sonometer.

Another method of analyzing sound is based upon the property of resonance. Instead of making the boxes like those adapted for tuning-forks, as described on page 283, they may be made spherical, with an opening on one side for the sound waves to enter and a small projection on the opposite side to be placed in the ear (Fig. 160). Such a resonator of a proper size will respond to only one given pitch, and a series of different sizes will enable one to detect the presence of overtones in sounds of musical instruments and of the voice. With such means it has been found that there are few sounds which are simple, that is, consist of only a single rate of vibration. A tuning-fork upon its resonance box is nearly so.

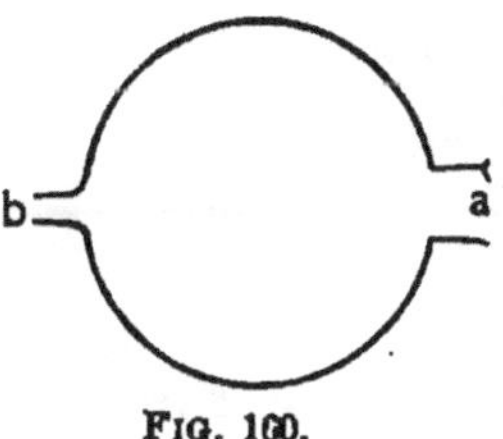

FIG. 160.

Most sounds are highly complex, and their several constituents usually stand in a certain numerical relation to each other, such that the first overtone above the fundamental or pitch is made by twice as many vibrations as the fundamental, the second by three

INSTRUMENTS.	1	2	3	4	5	6	7	8	9	10
ORGAN PIPES. Wide stopped	✓									
ORGAN PIPES. Narrow stopped	✓		✓		✓		✓		✓	
ORGAN PIPES. Narrow cylinder	✓	✓	✓	✓	✓	✓				
ORGAN PIPES. Principal (Wood)	✓	✓	✓							
ORGAN PIPES. Conically narrow at top,	✓				✓	✓	✓			
Flute	✓	✓	✓	✓						
Violin	✓	✓	✓	✓	✓	✓	✓	✓	✓	✓
Piano	✓	✓	✓	✓	✓	✓	✓	✓		
Bell	✓	✓	✓	✓	✓	✓	✓			
Clarionet	✓		✓		✓		✓		✓	
Bassoon	✓	✓	✓	✓	✓	✓	✓			
Oboë	✓	✓	✓	✓	✓	✓	✓			

FIG. 161.

times the fundamental, and so on; thus, if the fundamental or lowest sound is made by 100 vibrations, the first overtone will be made by 200, the second by 300, the third by 400, the fourth by 500, and so on, up to the twentieth and beyond. One hears these all together, and unless one has listened for them, they may never have been noticed, and they may have been thought of altogether as of a single sound. Listen to a distant church-bell, and two or three different sounds may be detected.

All the above-mentioned constituents are not present in all sounds, but when present they are in these ratios. Also, when present, some constituents are relatively stronger than others, and this difference in the number and strength of the overtones is what makes the difference in the character or quality of the sounds of musical instruments, such as the piano, the flute, and the violin.

The accompanying table gives the composition of the tones of a number of common musical instruments. The heavy lines in column 1 represent the fundamental sound or pitch of the instrument; the lighter ones in the other columns represent the overtones present. Thus in the flute, besides the fundamental, the second, third, and fourth in the series are present. The clarionet has for overtones the third, fifth, seventh, and ninth; and the violin has the whole series, though the eighth, ninth, and tenth are weaker sounds than the others, and are indicated by lighter lines.

The Voice.—No two voices are alike. All of us can distinguish the voices of our acquaintances, whether the persons can be seen or not, as readily as we can distinguish the sound of different musical instruments.

The vocal organs in man are highly complex in structure, made up of bones, cartilage, muscles, tubes, and cavities of variable size. The air enters the lungs through the mouth and nose cavities **M** and **C** (Fig. 162), thence through a special device called the larynx **G**, into a tube called the *trachea*, which branches into smaller tubes called *bronchial tubes* in the lungs. The sound-producing organ consists of two cartilages, which, like two lips, nearly close the opening to the trachea. These cartilages, which are called *vocal cords* can, at will, be made tense or loose, and also can separate to a greater or less degree so as to permit more or less air to pass either way. The vibrations of these vocal cords when air is forced through the orifice, give rise to the air vibrations; the degree of tension given to them determines the pitch, and the amount of air forced through them determines the loudness of the sound.

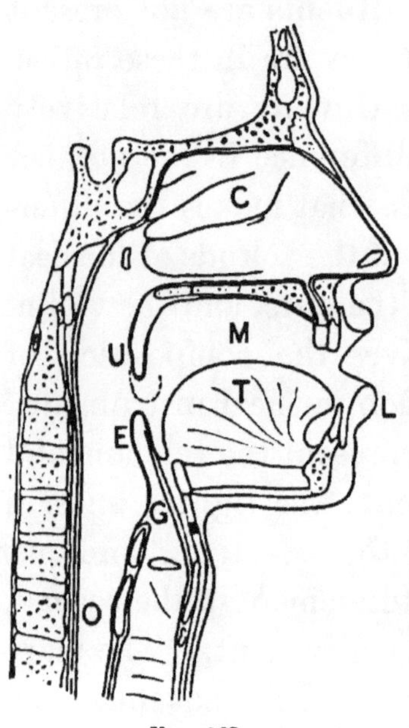

FIG. 162.

By stretching two pieces of thin rubber membrane over the end of a tube so as to leave a narrow fissure, as in Fig. 163, an artificial glottis can be made which, if blown through, will produce a sound, the pitch of which will depend upon how tightly the membrane is stretched.

FIG. 163.

The sound waves produced by the cords escape by way of the mouth **M** and nose cavity **C**, both of them acting as resonant cavities. The mouth can be greatly varied in size by the movements of the lower jaw **L** and tongue **T**, soft palate **U**, the cheeks, and the lips.

These changes in size greatly modify the resonance. In noting the movements of these when speaking the vowels *a e i o u*, one can perceive that the shape of the mouth cavity determines what sound can be made. One cannot speak the letter *a* when the mouth and tongue are in proper position for speaking *o*. The nasal cavity C has much to do in articulations, and when the lining membrane is swollen, as in head colds, or the nose is pinched so as to close the cavity, many words cannot be spoken plainly, especially such as require the sound of *m* or *ng*. Pinch the nostrils and try to say the word "something," and it will be discovered how important the resonating cavity of the nose is for articulation.

In different individuals the size, shape, texture, and muscular control of each of these differ so much as to give character to the sound of each voice. There are not less than a hundred different muscles called into action in talking, and so admirably are they adjusted to each other that they all work automatically as one complex machine, and no more require conscious supervision than does winking or walking.

The Opeidoscope. — The string telephone indicates the vibratory motions communicated to it by sound waves, but these may be made still more apparent by tying a piece of letter-paper over one end of a tube an inch or two in diameter and three or four inches long (Fig. 164), and gluing a bit of mirror a quarter of an inch square upon the middle of it. If light be directed upon this from the *porte-lumière* (p. 229), and the beam be reflected to the wall, ten or fifteen feet away, in a darkened room, any vibratory motion of the paper will be indicated by the movements of the beam of light. If sounds be made in the open end of the tube, various curves will be described on the wall, and by tilting the end with the mirror, complex wave figures will be formed, which will depend upon the pitch, the overtones, and the loudness of the sound. If a tune be tooted into it, the form will change for every different note, but will be always the same for any given one. This instrument is called the *opeidoscope*.

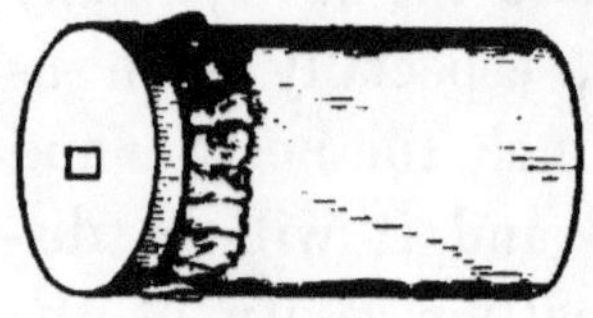

FIG. 164.

A soap film stretched across one end of a lamp chimney and held in the beam from a *porte-lumière* will reflect it to the wall. A lens with ten or twelve-inch focus will give an enlarged image of the film upon the screen. Prismatic colors may be seen, and if sounds be made near the open end of the tube, the film will be colored with beautiful coruscations, which will change with every pitch of sound made.

The Phonograph. — If in place of the mirror a short needle be fixed to the middle of the vibrating surface or diaphragm, as at **b**, Fig. 165, it will move in and out as often as the diaphragm does. If the point of the needle touches a surface of wax spread on the cylinder **c** that is being rotated, a series of indentations will be made in the wax corresponding exactly with the number of vibrations that the needle makes. If the cylinder is turned round so as to permit the needle to fall into the indentations at first made, its rotation will cause the needle, and with it the diaphragm to which it is attached, to make the same number and kind of vibrations that produced the indentations. It will, consequently, record and reproduce any kind of a sound. The characteristics of all musical instruments as well as of voices, may be preserved and reproduced at will. This machine is called the *phonograph.* Its operation is an illustration of what was said on page 280, that sound vibrations cause surfaces upon which they fall to vibrate in like manner as themselves. It also shows that in speech there is nothing but definite complex vibratory motion.

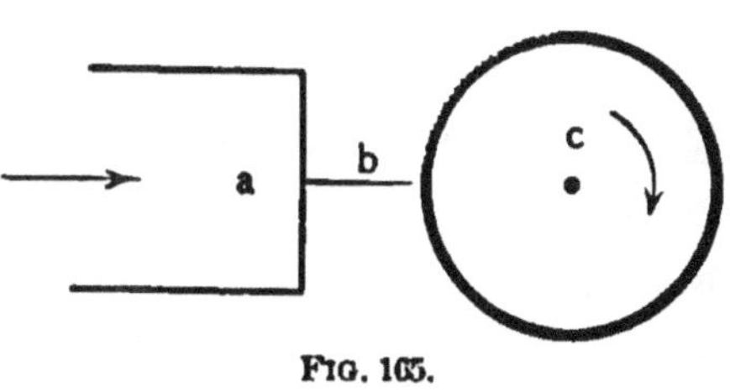
a
b
c
FIG. 165.

Music. — When a single sound is produced by the voice or upon any instrument, as a flute or piano, by striking a single key, that sound is called a *tone.* A number of tones of like quality, varying more or less in pitch, following each other with regularity, is called a *tune* or *melody.* Thus the notes written below constitute a melody, whether played upon some instrument or hummed by the voice.

There's no place like home, there's no place like home.
FIG. 166.

Whenever words are sung to the varying notes it is called a *song*. Upon many musical instruments it is impossible to produce more than one tone at a time, for example, the flute, the cornet; also with the voice. Other instruments permit several tones to be made upon them at the same time, as the violin, the harp, piano, and organ. It is a fact familiar to every one that some sounds when heard together are very disagreeable to the ear, yet when heard separately are agreeable enough, while other sounds when heard together are pleasing.

There are certain relations that are found to exist between these agreeable and disagreeable sounds, to understand which it is needful to be acquainted with the *musical scale*.

It is customary to write music on a group of five lines, called the *staff*. Each of the lines and spaces is designated by a letter that is constant.

The staff and letters are thus represented (Fig. 167). Only seven letters of the alphabet are used, and these

FIG. 167.

are repeated above and below as far as necessary. The musical alphabet begins with the letter *C* in the place indicated above, and if notes are placed upon the lines and spaces as shown, a series of tones will be represented called the *natural scale*, which is sometimes sung to the syllables *do*, *re*, *mi*, and so on. If the note *C* be sung or played at the same time as the note *D*, the result is a higher, unpleasant sound, which is called a *discord;* while if the note *G* is heard with the note *C*, the compound sound will be what is called a *concord* or *harmonious* sound. In like manner, the sound of the higher *C* with the lower *C* is tolerated by the ear. These sounds all differ from each other only in pitch.

In order that there shall be uniformity in pitch, musicians have adopted as a standard for the letter *C* on the added line below the staff 261 vibrations per second. This *C* is the middle *C* of the piano.

Number	Name	Length of String	Ratio of each to First
1	Do	36	
2	Re	32	$\frac{8}{9}$
3	Mi	28.8	$\frac{4}{5}$
4	Fa	27	$\frac{3}{4}$
5	Sol	24	$\frac{2}{3}$
6	La	21.6	$\frac{3}{5}$
7	Si	19.2	$\frac{8}{15}$
8	Do	18	$\frac{1}{2}$

Let the sonometer have both strings tuned to the same pitch. Calling the tone given by one string *do*, place the movable bridge under the second string near one end, and move it slowly towards the other end until, when plucked, it gives the tone *re* as accurately as can be judged by the ear.

Measure the length of both strings. The shorter one will be found to be $\frac{8}{9}$ the length of the other. Do the same for each letter, and make a table of the measures and results as above.

In this table the assumed length of the string giving the note *do* is 36 inches, in which case the length giving *re* will be 32 inches, that giving *sol*, 24 inches, and so on as given. The ratio of 32 to 36 is as 8 to 9, — the length of the second string is $\frac{8}{9}$ the length of the first. The ratio of 24 to 36 is as 2 to 3, — the length of string for *sol* is $\frac{2}{3}$ that for *do;* and so on for each of the others. As the vibratory rates of strings are inversely as their lengths, it follows that the vibratory rate of *re* is $\frac{9}{8}$ that of *do;* and of *sol*, $\frac{3}{2}$ that of *do*. The fraction representing the ratio of the length of one string to the first one will, when inverted, represent their relative number of vibrations, or while *do* makes 8 vibrations, *re* makes 9, and while *do* makes 2 vibrations, *sol* makes 3.

If, therefore, the fundamental *do* be tuned to standard pitch of 256 vibrations, *re* will make $256 \times \frac{9}{8} = 288$, and *sol* will make $256 \times \frac{3}{2} = 384$ in a second. Computing the values of each of the numbers of the scale as above, one will have another table which will represent the number of vibrations for each letter of the scale.

LETTER	RATIO	
C		256
D	$\frac{9}{8}$	288
E	$\frac{5}{4}$	320
F	$\frac{4}{3}$	$341\frac{1}{3}$
G	$\frac{3}{2}$	384
A	$\frac{5}{3}$	$426\frac{2}{3}$
B	$\frac{15}{8}$	480
C	$\frac{2}{1}$	512

These ratios and numbers may be verified in another way. Choose two glass tubes six or eight inches long and half an inch in diameter. Stop one end of one with a cork and blow across the open end; it will give a sound of definite pitch. Fit a cork to the other tube, such as will fit snugly and can be pushed through it at will. Adjust this cork so that the tube, when blown across, will give the same pitch as the first tube; if the first gives *do*, move the cork in the second tube so it will give the note *re;* measure the lengths of both tubes and compare them. The shorter one will be found to be $\frac{8}{9}$ the length of the longer. In like manner, each of the other notes of the scale can be sounded by shortening to the proper length the column of air in the tube with the movable plug; the ratios of their lengths to that of the first tube will give the same figures as those given by the sonometer.

Such a series of fractions as those given above are called *musical ratios.*

Interference of Sound Waves. — Strike a small tuning-fork, and, holding it near the ear, roll the stem between the thumb and finger. A decided change in the strength of the sound will be noticed, and if the fork be carefully tuned, a position may be found where the sound can scarcely be heard, and a slight turn from that position in either direction will make the sound again audible. This may be understood by remembering that the two prongs of the fork beat towards and away from each other while vibrating, hence each of them will be generating a rarefaction in the air between the prongs at the same time they are condensing the air in front of them. But the rarefaction can be conducted away only past the prongs, then it begins to be distributed in every direction. At the corners the rarefaction meets the condensation, and the two neutralize each other because their motions are in opposite directions. In certain directions these balance each other so there is no wave motion at all, and consequently no sound. This phenomenon is called *interference.*

When two sounds are nearly, but not quite, of the same pitch, they produce what are called *beats*, which may be easily heard from piano-strings not quite in tune, and from two similar tuning-forks, if one has a bit of wax stuck to its prongs. If one sound is produced by 256 vibrations a second, and the other by 257, the combined sound will be stronger once a second, and the beats can be heard without difficulty; if they differ by two or three vibrations per second, there will be heard two or three beats per second. The greater the difference, the more frequent the beats, and presently the beats produced in this way quite spoil the tones. If, however, the beats occur as many as twenty or more times a second, they give a continuous sound of definite pitch. If the number of beats per second be a submultiple, such as one-half or two-thirds the number of vibrations of either of the tones that produce them, they add an agreeable constituent to the sound, and altogether they constitute a *chord*. Thus in the case of the two notes *C* and *G*, when the former makes 256 vibrations and the latter makes 384, — the ratio being 2 to 3, — the number of beats per second will be the difference between the two numbers, namely, 128, which is one-half of 256. The beats coincide with the fundamental 128 times a second. The three heard together are as if a third tone an octave lower had been added. When *C* and *D* are sounded together, *C* makes 256 vibrations and *D* 288, and there are 32 beats per second. The compound is disagreeable to the ear and is called *discordant*.

If more than two sounds be produced together they

must all be in some simple ratio to each other if they are to be pleasing. This is the case when the notes of what is called the *common chord* are sounded together, in which we have the ratio $CE, \frac{5}{4} : CG, \frac{3}{2} : EG, \frac{6}{5}$. If more notes be added they must be duplicates of these in some other octave.

Harmony. — When music is thus written to be played or sung it is called *harmony*, and throughout all the changes of pitch in each part the simple ratios

AMERICA.

FIG. 168.

given are maintained, as may be seen by taking any four-part music, substituting the vibration numbers for the notes, putting them in fractional form, and reducing them to their lowest terms.

In the above music, written in two parts, treble and bass, the ratios as expressed by the fractions are written out for the notes sounded together. Observe that the numbers are all such as have been described as derived from the measurement of vibration numbers, and are essential for harmony. The ratio 2:3 occurs eight times, 5:3 and 4:3 each five times, and 5:4 six times, while such as 16:9 occurs but twice, and 64:45 only once. If other parts were compared, similar ratios would appear. One may thus verify these music relations with such tunes as "Auld Lang Syne" or "Greenville," when arranged in two or more parts, and assure himself that part music and all harmony consists of sounds whose vibration numbers are in simple ratios, and that any departure from these ratios is not endured by the ear, and has to be quickly followed by notes with the required relations.

CHANGE OF KEY.

Sharps and Flats. — The natural scale which we sing or play (p. 298) is not adapted to all pieces of music. One may easily satisfy himself on this point by singing "America" on the pitch of *C*. It will be found too low, and some higher pitch must be chosen. When a different pitch is taken the same ratios must be kept, and we do keep them when we sing *do*, *re*, *mi* of the scale.

On the staff is written the actual vibration numbers of the various notes opposite the notes themselves. If we choose the pitch of *G* for the new scale, calling that note *do*, and calculate the vibration numbers for the new scale, they all will be found to agree so nearly with the numbers calculated for the key of *C* (except the note on *F* on the upper line) that they may be used for musical purposes. The note on *F* for the key of *C* has 682 vibrations ($512 \times \frac{4}{3}$), while for the key of *G* it should have 720 ($384 \times \frac{15}{8}$). This number, 682, is considerably too low for the proper pitch, and it

Fig. 169.

becomes necessary to interpose a new note between *F* and *G*. As the new note takes the place of *F* and is higher in pitch, it is called *F sharp*, and is marked by the symbol ♯ to indicate it.

If the pitch of *F* is chosen for the new scale and the vibration numbers for *C*, 256, be compared with the numbers computed for *F*, 341, they will be found to coincide, with the exception of one on the letter *B*. In the natural scale *B* makes $256 \times \frac{15}{8} = 480$ vibrations, but the fourth in the scale of *F* should make $341 \times \frac{4}{3} = 454$ vibrations. A new note of lower pitch must be substituted for *B* and we call it *B flat*. Thus by using any one of the vibration numbers of the natural scale of *C* as a basis, and multiplying that number by the series of fractions by which the numbers of the scale of *C* were obtained, and comparing these numbers with those of the scale of *C*, one may see at once which letters need to be modified. If too low, a sharp (♯) is employed; if too high, a flat (♭) is used. Scales built upon other tones than *C* receive the name of the tone they are built upon, as the key of *G sharp*, the key of *B flat*.

Harmonic Interference. — It has been stated that sounds are rarely simple, but are made up of a series of sounds which are multiples of the lowest or fundamental sound. If we have different fundamentals sounding together, then overtones must be present to strengthen or to weaken each other by interference. Thus, if the note *C* (bass) upon a piano be struck, the overtones belonging to it are indicated by the black notes upon the staff. *E* and *G* in like manner have their overtones represented. If the *C* and *E* be compared, the third overtone of the *E* coincides with the fourth of the *C*, and the first and third of the *G* with the second and fifth of *C*. All the rest are more or less interfering and discordant. If it were not for the fact that the higher overtones are generally feeble, the most of that which we now call music would be unendurable because of the discordances which are really present.

FIG. 170.

Consider what a number of sounds there must be when a full organ is played with the harmony. There may be as many as fifty or more series of pipes, each with eight or more pipes sounding at one time. These would give as many as four hundred fundamental sounds, while the overtones must be reckoned by thousands. With a full orchestra with twelve or fifteen different kinds of instruments, the overtones would be in greater variety than on the organ. The vocal organs are like a musical instrument, and the overtones are not alike in pitch or intensity for any two individuals. We therefore hear a wonderful variety of sounds when a hundred or more persons are singing together. The interfering sounds and high overtones do not have sufficient intensity to be heard very far, hence the richness of many blended voices when heard at the distance of two or three hundred feet or farther. Such sounds as are produced by many persons talking at once, by carriages when in motion, by tin horns and tin pans, all of which are called noises, are simply compound sounds without definite ratios to each other, and hence are discordant.

The Ear. — Thus far the phenomena of sound have been considered as purely physical, the origin and distribution of waves in the air and in other elastic bodies alone being regarded. Such phenomena could exist and be studied if there were no such things as ears, just as the physical phenomena of the ether waves could be studied without eyes. Eyes and ears are helpful in either study, but both are not essential. Physical phenomena and sensation are so utterly different from each other that they cannot be compared. A needle-point can be thrust into a piece of wood or into the finger, the physical phenomena are the same; but in the finger there is a sensation of pain that accompanies the action, but is not a part of it. Sensation implies a nervous system and a sensorium, that is, a seat of consciousness.

Between the sound vibrations in the air and the sensorium in the brain there is a complicated piece of apparatus called the *ear*. In man the external part, which is called the *pinna*, is not an important part for hearing, for it may be quite removed without making any appreciable difference in the perception of sounds. The inner part is a series of tubes and chambers within the solid bony structure of the skull. A small tube **A**, called the *auditory canal*, opens inward to the depth of about an inch and a quarter. The inner end of this canal is closed by a membrane, called the *tympanic membrane*, stretched across it at an angle. It is curved inward at its middle part as if pulled from within. Behind this membrane is an air chamber **B**, which has a tube **C**, called the *eustachian tube*, leading to the back of the mouth. In this chamber are three small bones: one attached to the middle part of the membrane, the second to the first, and the third, which has the form of a stirrup, is made fast to the second on one side and to another membrane on the side opposite. This second membrane closes a complicated cavity, the *bony laby-*

FIG. 171.

rinth in the bone of the skull. This labyrinth is filled with fluid in which floats a *membranous labyrinth* having several distinct parts. On one side of the sac are three *semicircular canals* **D**, and on the other a coiled tube **E**, called the *cochlea*. The membranous labyrinth in turn is filled with fluid containing a great number of minute crystals called *otoliths*. At certain places inside the sac, canals, and cochlea, are small bodies known as auditory cells, each bearing small hairs on its surface. The structure is extremely complicated in the cochlea, where there is an apparatus known as the *organ of Corti*, made up of over three thousand sets of cells, each set consisting of five or more cells. The auditory nerve **G**, coming from the brain breaks up into branches going to each group of cells and again branching, so that every cell has a twig. There are not less than 20,000 of these fibres of Corti in each human ear.

One may now trace the physical actions within the ear. The air vibrations in the auditory canal make the tympanic membrane vibrate; this in turn shakes the series of bones, and the stirrup shakes the window membrane to the labyrinth. This shakes the contained liquid, otoliths, fibres, and nerve cells, and the disturbance of the nerve ends is transmitted by the nerve to the seat of sensation. *Then* we hear. If any part of this mechanism is absent or impaired, we become deaf. Thus, if the tympanic membrane is ruptured or thickened so it cannot vibrate, deafness is the result. Sound vibrations communicated to the bones of the head may be heard, provided the interior parts of the ear are in order. A tuning-fork may be heard by touching its stem to the skull or teeth; and a sheet of thin wood or metal held by the edge between the teeth enables some persons to hear better. Such a device is called an *audiphone.*

From the auditory canal to the auditory nerve there is simply a series of vibratory movements, which become weaker and weaker because more and more diffused, until in the parts of the labyrinth they can be only of molecular dimensions, that is, only millionths of an inch. They become changed in magnitude, not in character. The particular functions of these various parts of the ear are but partially known. The fibres of Corti were thought for a time to be sympathetic, so that each one could respond to some particular rate of vibration; but that view is now believed to be unfounded, and at present there is no explanation of what goes on in the labyrinth. Movements of some sort are traceable to the base of the brain; there they are interpreted as sounds. How physical actions of any kind can produce sensations is unimaginable.

Limits of Audibility. — For the production of high musical tones, sets of steel rods are used that have definite lengths and diameters. A steel rod 2.6 inches long and .78 of an inch in diameter will give 20,000 vibrations per second. As the rods are made shorter the pitch becomes higher, so that one 1.8 inches long gives 40,000 vibrations. Most persons fail to hear any sound of 30,000 vibrations per second, but a few can hear as many as 35,000 or upwards. It is altogether probable that this limit for any one is due only to the difficulty there is in giving energy enough to the higher sounds, rather than that the apparatus for hearing is itself limited. One is not called deaf because he cannot hear a weak sound, if by making the sound stronger he can hear it.

All sounds involve energy in the mechanical sense, so that all problems of sound are problems of the transference and transformation of energy.

Electrical Resistance, Diameter, Cross-Section, etc., of Copper Wire, American Gauge, Temperature 24° C.

Gauge No.	SIZE.		WEIGHT.		RESISTANCE.			Capacity in Amperes.
	Diam. in In.	Area in Sq. In.	Lbs. per 1000 Ft.	Feet per Lb.	Ohms per 1000 Ft.	Feet per Ohm.	Ohms per Lb.	
0000	.4600	.166191	639.60	1.564	0.051	19929.7	.0000785	312
000	.4096	.131790	507.22	1.971	0.063	15804.9	.000125	262
00	.3648	.104590	402.25	2.486	0.080	12534.2	.000198	220
0	.3249	.082932	319.17	3.133	0.101	9945.3	.000315	185
1	.2893	.065733	252.98	3.952	0.127	7882.8	.000501	156
2	.2576	.052130	200.63	4.994	0.160	6251.4	.000799	131
3	.2294	.041339	159.09	6.285	0.202	4957.3	.001268	110
4	.2043	.032784	126.17	7.925	0.254	3931.6	.002016	92.3
5	.1819	.025998	100.05	9.995	0.321	3117.8	.003206	77.6
6	.1620	.020617	79.34	12.604	0.404	2472.4	.005098	65.2
7	.1443	.016349	62.92	15.893	0.509	1960.6	.008106	54.8
8	.1285	.012766	49.90	20.040	0.643	1555.0	.01289	46.1
9	.1144	.010284	39.58	25.265	0.811	1233 3	.02048	38.7
10	.1014	.008153	31.38	31.867	1.023	977.8	.03259	32.5
12	.0808	.005129	19.74	50.659	1.626	615.02	.08237	23.
14	.0641	.003147	12.41	80.580	2.585	386.80	.20830	16.2
16	.0508	.002029	7.81	128.041	4.582	243.25	.52638	11.5
18	.0403	.001276	4.91	203.666	6.536	152.99	1.3312	8.1
20	.0320	.000802	3.086	324.045	10.394	96.21	3.3438	5.7
25	.0179	.000252	0.967	1034.126	33.135	30.18	34.298	2.4
28	.0126	.000125	0.484	2066.116	66.445	15.05	137.283	1.4
30	.0100	.000079	0.302	3311.258	105.641	9.466	349.805	1.0
32	.0079	.000050	0.190	5263.158	168.011	5.952	884.267	0.70
34	.0063	.000031	0.121	8264.463	267.165	3.743	2207.98	0.50
36	.0050	.000020	0.075	13333.33	424.65	2.355	5661.71	0.35
40	.0031	.000008	0.030	33333.33	1074.11	0.931	35803.8	0.17

Mechanical, Heat, and Electrical Equivalents.

1 Foot-pound =	.0000303	Horse-power per minute.
	.001818	" " " second.
	.0003767	Watt per hour.
	.0226	" " minute.
	1.356	" " second.
	.001288	Heat Unit.
1 Horse-power =	746	Watts.
	1.980000	Foot-pounds per hour.
	33000	" " " minute.
	550	" " " second.
	2550	Heat Units per hour.
	42.53	" " " minute.
	.707	" " " second.
1 Heat Unit =	778	Foot-pounds.
	.00039	Horse-power per hour.
	.0235	" " " minute.
	1.41	" " " second.
	0.29083	Watt per hour.
	17.45	" " minute.
	1046.98	" " second.
1 Watt =	2654.4	Foot-pounds per hour.
	44.239	" " " minute.
	.737	" " " second.
	.0013406	Horse-power.
	3.4384	Heat Units per hour.
	.0573	" " " minute.
	.000955	" " " second.

INDEX.

ADVERTISEMENTS

NATURAL SCIENCE TEXT-BOOKS.

www.ingramcontent.com/pod-product-compliance
Lightning Source LLC
Chambersburg PA
CBHW021120270326
41929CB00009B/963

* 9 7 8 3 7 4 4 7 4 9 9 8 5 *